I0502850

Prepared in cooperation with the U.S. Army Corps of Engineers

Grain-Size Distribution and Selected Major and Trace Element Concentrations in Bed-Sediment Cores from the Lower Granite Reservoir and Snake and Clearwater Rivers, Eastern Washington and Northern Idaho, 2010

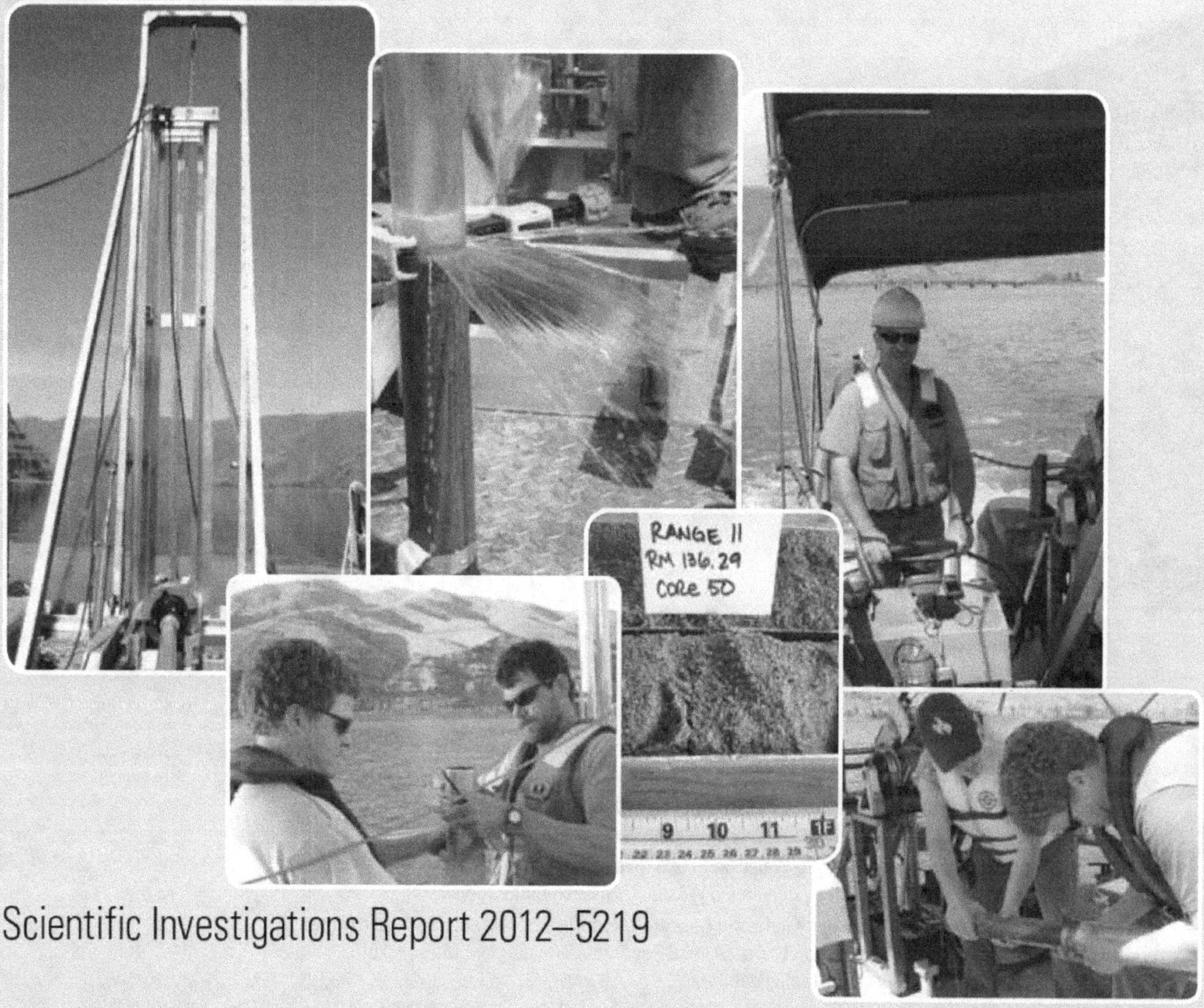

Scientific Investigations Report 2012–5219

U.S. Department of the Interior
U.S. Geological Survey

Grain-Size Distribution and Selected Major and Trace Element Concentrations in Bed-Sediment Cores from the Lower Granite Reservoir and Snake and Clearwater Rivers, Eastern Washington and Northern Idaho, 2010

By Christopher L. Braun, Jennifer T. Wilson, Peter C. Van Metre,
Rhonda J. Weakland, Ryan L. Fosness, and Marshall L. Williams

Prepared in cooperation with the U.S. Army Corps of Engineers

Scientific Investigations Report 2012–5219

U.S. Department of the Interior
U.S. Geological Survey

U.S. Department of the Interior
KEN SALAZAR, Secretary

U.S. Geological Survey
Marcia K. McNutt, Director

U.S. Geological Survey, Reston, Virginia: 2012

This and other USGS information products are available at http://store.usgs.gov/

U.S. Geological Survey
Box 25286, Denver Federal Center
Denver, CO 80225

To learn about the USGS and its information products visit http://www.usgs.gov/
1-888-ASK-USGS

Suggested citation:
Braun, C.L., Wilson, J.T., Van Metre, P.C., Weakland, R.J., Fosness, R.L., and Williams, M.L., 2012, Grain-size distribution and selected major and trace element concentrations in bed-sediment cores from the Lower Granite Reservoir and Snake and Clearwater Rivers, eastern Washington and northern Idaho, 2010: U.S. Geological Survey Scientific Investigations Report 2012–5219, 81 p.

Contents

Figures

Table

Conversion Factors

SI to Inch/Pound

Multiply	By	To obtain
Length		
centimeter (cm)	0.3937	inch (in.)
millimeter (mm)	0.03937	inch (in.)
meter (m)	3.281	foot (ft)
kilometer (km)	0.6214	mile (mi)
meter (m)	1.094	yard (yd)

Inch/Pound to SI

Multiply	By	To obtain
Length		
inch (in.)	2.54	centimeter (cm)
inch (in.)	25.4	millimeter (mm)
foot (ft)	0.3048	meter (m)
mile (mi)	1.609	kilometer (km)
Volume		
cubic yard (yd^3)	0.7646	cubic meter (m^3)
Flow rate		
cubic foot per second (ft^3/s)	0.02832	cubic meter per second (m^3/s)

Vertical coordinate information is referenced to the North American Vertical Datum of 1988 (NAVD 88).

Horizontal coordinate information is referenced to North American Datum of 1983 (NAD 83).

Grain-Size Distribution and Selected Major and Trace Element Concentrations in Bed-Sediment Cores from the Lower Granite Reservoir and Snake and Clearwater Rivers, Eastern Washington and Northern Idaho, 2010

By Christopher L. Braun, Jennifer T. Wilson, Peter C. Van Metre, Rhonda J. Weakland, Ryan L. Fosness, and Marshall L. Williams

Abstract

Lower Granite Dam impounds the Snake and Clearwater Rivers in eastern Washington and northern Idaho, forming Lower Granite Reservoir. Since 1975, the U.S. Army Corps of Engineers has dredged sediment from the Lower Granite Reservoir and the Snake and Clearwater Rivers in eastern Washington and northern Idaho to keep navigation channels clear and to maintain the flow capacity. In recent years, other Federal agencies, Native American governments, and special interest groups have questioned the negative effects that dredging might have on threatened or endangered species. To help address these concerns, the U.S. Geological Survey, in cooperation with the U.S. Army Corps of Engineers, collected and analyzed bed-sediment core samples (hereinafter cores) in Lower Granite Reservoir and impounded or backwater affected parts of the Snake and Clearwater Rivers. Cores were collected during the spring and fall of 2010 from submerged sampling locations in the Lower Granite Reservoir, and Snake and Clearwater Rivers. A total of 69 cores were collected by using one or more of the following corers: piston, gravity, vibrating, or box. From these 69 cores, 185 subsamples were removed and submitted for grain size analyses, 50 of which were surficial-sediment subsamples. Fifty subsamples were also submitted for major and trace elemental analyses. Surficial-sediment subsamples from cores collected from sites at the lower end of the reservoir near the dam, where stream velocities are lower, generally had the largest percentages of silt and clay (more than 80 percent). Conversely, all of the surficial-sediment subsamples collected from sites in the Snake River had less than 20 percent silt and clay. Most of the surficial-sediment subsamples collected from sites in the Clearwater River contained less than 40 percent silt and clay. Surficial-sediment subsamples collected near midchannel at the confluence generally had more silt and clay than most surficial-sediment subsamples collected from sites on the Snake and Clearwater Rivers or even sites further downstream in Lower Granite Reservoir. Two cores collected at the confluence and all three cores collected on the Clearwater River immediately upstream from the confluence were extracted from a thick sediment deposit as shown by the cross section generated from the bathymetric surveys. The thick sediment deposits at the confluence and on the Clearwater River may be associated with floods in 1996 and 1997 on the Clearwater River.

Fifty subsamples from 15 cores were analyzed for major and trace elements. Concentrations of trace elements were low, with respect to sediment quality guidelines, in most cores. Typically, major and trace element concentrations were lower in the subsamples collected from the Snake River compared to those collected from the Clearwater River, the confluence of the Snake and Clearwater Rivers, and Lower Granite Reservoir. Generally, lower concentrations of major and trace elements were associated with coarser sediments (larger than 0.0625 millimeter) and higher concentrations of major and trace elements were associated with finer sediments (smaller than 0.0625 millimeter).

Introduction

The Lower Granite Dam impounds the Snake and Clearwater Rivers in eastern Washington and northern Idaho. Backwater from Lower Granite Reservoir extends to just upstream from the confluence of the Snake and Clearwater Rivers. The Snake and Clearwater Rivers transition at their confluence from free-flowing water to backwater caused by the Lower Granite Dam, and backwater marks the upstream extent of the reservoir pool. Delta deposits of bed-sediment material form as velocity and transport capacity diminish where streams enter the reservoir pool. In addition to deltaic deposition of primarily coarse sediments, processes in reservoirs include deposition of fine sediments from homogeneous flow, and transport and deposition of sediment from stratified flow (Fan and Morris, 1992, p. 355). Since its impoundment in 1975, 2.6 million cubic yards (2.0 million cubic meters) of sediment have been deposited annually into Lower Granite Reservoir and the Snake and Clearwater Rivers that flow into the reservoir (U.S. Army Corps of Engineers, 2003). The Snake River continues downstream from Lower Granite Dam; upstream from the confluence of the Snake and Columbia Rivers, the USACE operates three additional dams on the Snake River as part of the Snake River System (U.S. Army Corps of Engineers, 2002) (fig. 1). Historically, the USACE dredged sediment from the Snake River System, including Lower Granite Reservoir, to keep navigation channels clear and to maintain the flow capacity. Increases in reservoir stage in Lower Granite Reservoir caused by sedimentation also reduce the effectiveness of the levees protecting Clarkston, Wash., and Lewiston, Idaho, from flooding (Greg Teasdale, U.S. Army Corps of Engineers, written commun., 2011). In recent years, other Federal agencies, Native American governments, and special interest groups have questioned the negative effects that dredging might have on threatened or endangered species. The negative effects of dredging might include biological effects (on distribution, behavior, migration, feeding, spawning, development, and fish injury), physical effects (on disturbance, displacement, avoidance, entrainment, burial, noise, sedimentation, turbidity, suspended sediments, and habitat and food source modification), and water-quality effects (on acute toxicity, bioavailability, bioaccumulation, and exposure pathways) (U.S. Army Corps of Engineers, 2004).

To address these concerns, the USACE initiated a multiyear project to assess the current status of sediment deposition in the reservoir and to explore alternative sediment control measures. The multiyear project included surveys of the sedimentary structures (bedforms) of Lower Granite Reservoir. Bed-sediment core samples (hereinafter cores) were collected by the U.S. Geological Survey in cooperation with the USACE during the spring and fall of 2010. A multibeam echosounding (MBES) bathymetric survey during fall 2009 and winter 2010 and an underwater video map (UVM) survey of sediment facies during fall 2009 and winter 2010, also part of the multiyear study (Williams and others, 2012), were used to help interpret surficial sedimentary structures (bedforms) in the study area. The MBES and UVM surveys are described in detail by Williams and others (2012). The data from all three surveys will be used to model flood hydraulics and sediment transport and to make biological assessments in support of the USACE Programmatic Sediment Management Plan (U.S. Army Corps of Engineers, 2003).

Purpose and Scope

This report describes the results from cores collected in the Lower Granite Reservoir and the Snake and Clearwater Rivers impounded by the reservoir during spring 2010 and fall 2010. Specifically, the grain-size distribution of surficial-sediment subsamples from cores collected underwater in Lower Granite Reservoir, the Snake and Clearwater Rivers, and the confluence of these two rivers are described, along with the down-core grain-size distribution in cores collected

at or near the confluence of the Snake and Clearwater Rivers. Grain-size analyzes are compared with results from previous surveys to assess the predominant surficial sediment grain-size classes and how they relate to bed-sediment accumulation history, surficial sedimentary structures (bedforms), and embeddedness (degree to which gravel, cobble, boulders, or snags are sunken into the silt, sand or clay of a river or lake bottom) in Lower Granite Reservoir, the Snake and Clearwater Rivers, and the confluence of these two rivers. Results from the grain-size analyses were used to provide a quantitative mechanism for verification of the facies map generated from the UVM surveys done by Williams and others (2012). Selected major and trace element concentrations measured in subsamples of cores also are discussed.

Previous Studies

Multibeam Echosounding Bathymetric Survey

During fall 2009 and winter 2010, the U.S. Geological Survey (USGS), in cooperation with the USACE, conducted a hydrographic survey using a MBES to develop a digital elevation dataset on 12 river miles (RM) of Lower Granite Reservoir and the Snake River, and 2 RM of the Clearwater River upstream from the confluence with the Snake River (figs. 1 and 2) (Williams and others, 2012). The confluence of the Snake and Clearwater Rivers is upstream from the Lower Granite Dam and is where the rivers transition from free-flowing to backwater. Data from the survey will be used by the USACE to better understand and predict sediment transport and deposition in the reservoir as part of its Programmatic Sediment Management Plan (U.S. Army Corps of Engineers, 2003). The digital elevation dataset also can be used to display river-bed elevation and geomorphology such as scour holes, rock outcrops, and bedforms like ripples and dunes. This survey represents a snapshot-in-time of benthic geomorphology that can rapidly change because of fluctuations in reservoir stage, river discharge, and boat traffic.

The MBES bathymetric survey was conducted from RM 130 to RM 142 on the Lower Granite Reservoir and Snake River and from RM 0 to RM 2 on the Clearwater River (fig. 2A). The survey mapped the full width of the river except areas along banks that were inaccessible to the boat or too shallow (less than 10 meters) to be measured with echosounding equipment. The survey was conducted in 1-mile segments, and the resulting datasets were composited to provide a continuous digital elevation dataset of the reservoir and rivers. The primary purpose for these data is to support the USACE's sediment transport modeling effort. However, these data also can provide a visual representation of bed geomorphology, which may be used for habitat assessments and other purposes. Figure 2B shows a two-dimensional overhead view of part of the surveyed area, which includes areas of sand dunes, a scour hole, basalt outcrop, and areas where bed material was removed for levee construction.

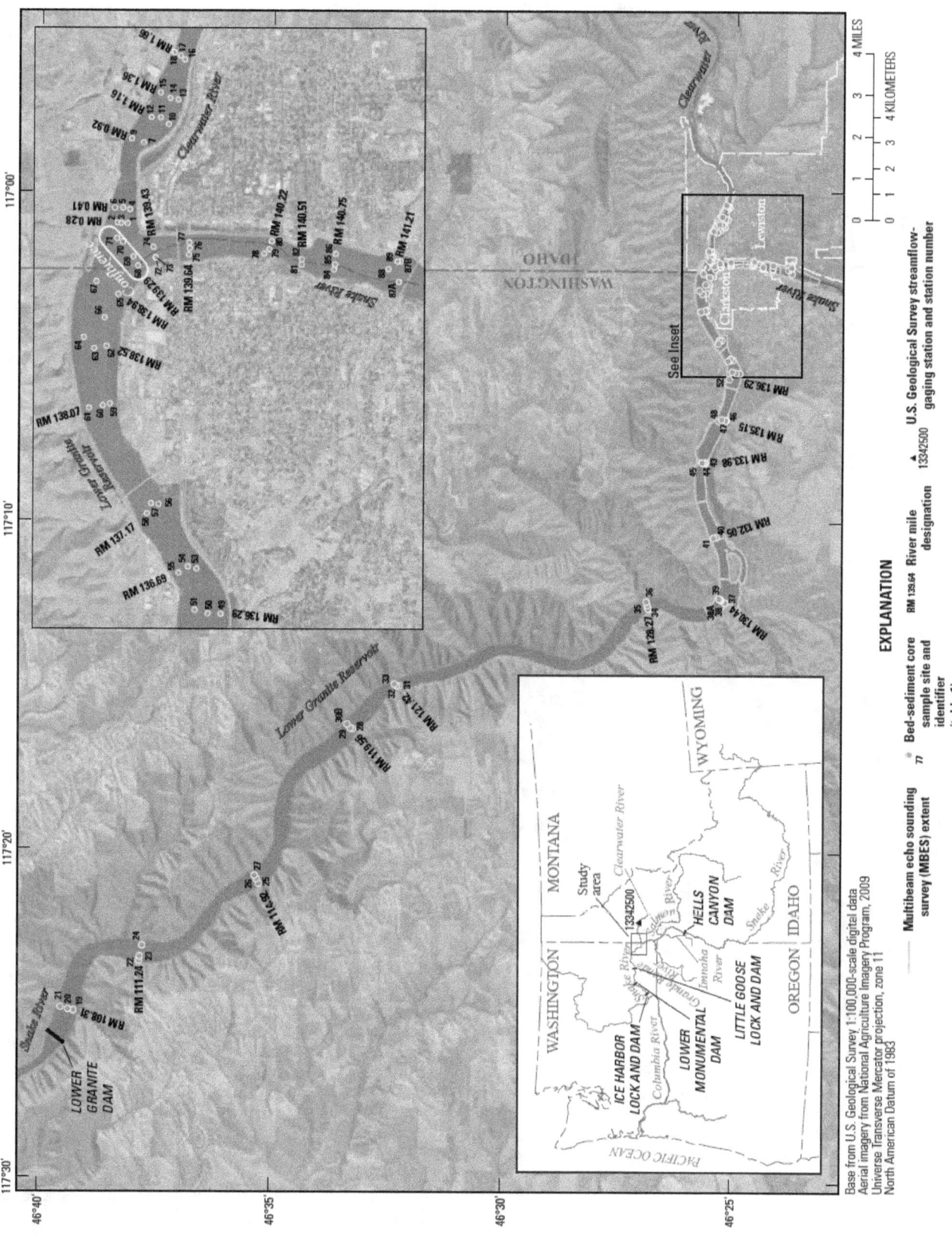

Figure 1. Study area and sediment coring locations in Lower Granite Reservoir and the Snake and Clearwater Rivers, eastern Washington and northern Idaho, 2010.

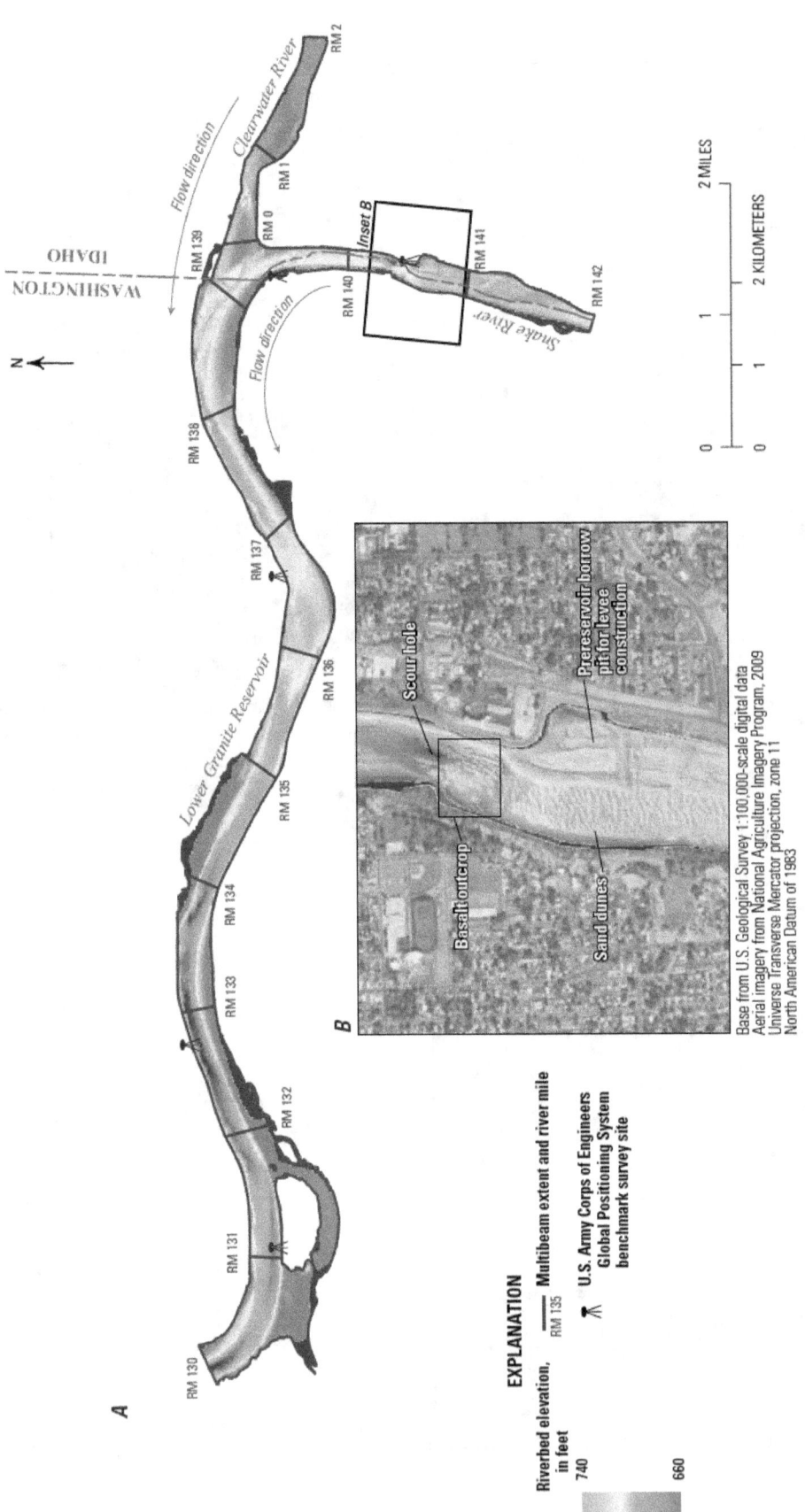

Base from U.S. Geological Survey 1:100,000-scale digital data
Aerial imagery from National Agriculture Imagery Program, 2009
Universe Transverse Mercator projection, zone 11
North American Datum of 1983

Figure 2.   *A,* Bathymetry from river mile 130 to river mile 142 of Lower Granite Reservoir and the Snake River and from river mile 0 to river mile 2 of the Clearwater River, eastern Washington and northern Idaho, and *B,* multibeam bathymetry overlying imagery showing geomorphologic features such as sand dunes, a scour hole, and rock outcrop between river miles 140 and 142 (modified from Williams and others, 2012, p. 6).

Underwater Video Map Survey

UVM surveys were done by the USGS in cooperation with the USACE during fall 2009 and winter 2010 (Williams and others, 2012) to visually identify the surface substrate type and percent embeddedness by determining particle size and the extent to which coarse substrate is surrounded or covered by silt and clay particles less than 0.0625 mm in diameter (commonly referred to as "fines" [Guy, 1969]). UVM surveys provide georeferenced information about the type and size of sediment on the surface of the bed. This information was used to enhance the bathymetric data and create a surficial sediment facies map of the current sediment distribution within the study area. The sediment facies map provides information on benthic habitat characteristics, variability of surface substrate, sediment transition zones, and unrealized areas of preferred species-specific habitat.

The UVM surveys were done between RM 107.73 and RM 141.78 on Lower Granite Reservoir and the Snake River and between RM 0.28 and RM 1.66 on the Clearwater River and at RM 139.29 at the confluence of the Snake and Clearwater Rivers. More than 900 video clips were recorded at discrete, equal-width increments (video points, fig. 3A) along 61 historic USACE survey lines and 5 existing longitudinal dredge-material deposit sites established by the USACE (fig. 3B). Video clips were recorded using a high-resolution, color-video camera outfitted with two high-power laser pointers that had a constant 4-inch (102 millimeter) separation as a scale of reference to determine sediment grain size (fig. 3A). The camera was lowered through the water column until it was close to the sediment-bed surface; sediment size and percent embeddedness were estimated based on a visual inspection of the video by an experienced hydrologist.

Sediment sizes and categories were defined as bedrock, boulder, cobble, gravel, sand, rip-rap, and fines (silt and clay). Coarse substrates (boulder, cobble, and gravel) were combined into one category for the purposes of this study because the occurrence of these types of substrates can vary greatly over a small area (fig. 3D).

Historically, facies maps were created by collecting sediment samples from the bed of the water body and analyzing the samples for grain size and sediment type. The level of effort and cost for such an endeavor limits the number of samples that can be collected, resulting in the creation of a facies map with a large amount of interpolation. Coupling results from the UVM survey with results from the MBES bathymetric survey enables the creation of a high resolution facies map (fig. 3) superior to a map created using traditional methods. The video record captures the primary attributes of the substrate, such as bottom type, texture, small bedforms, disturbance indicators, unusual features, and embeddedness; it also provides verification of the sediment type identified from the MBES data. For bedforms that are greater than the camera's field of view, especially in limited light conditions, the MBES bed-elevation data provide the fidelity to define larger geomorphological features. The bathymetric survey results can be used to identify bedform features that might otherwise be misinterpreted, on the basis of video analysis alone, such as a featureless bottom (Williams and others, 2012).

Methods

Core samples were collected at 67 locations (at each of two locations, 2 core samples were collected) in Lower Granite Reservoir, the Snake and Clearwater Rivers, and the confluence of these two rivers using one or more of the following corers: piston, gravity, vibrating (also called a vibracore), or box. From these 69 cores, 185 subsamples were removed and submitted for grain size analyses, 50 of which were surficial-sediment subsamples. Surficial samples were not submitted for grain-size analysis from cores that had surficial layers made up of predominantly organic matter and detritus that were unsuitable for grain-size analysis. A total of 50 subsamples were also submitted for major and trace elemental analyses. The cores were collected in April, May, and October 2010 between RM 0.28 and 1.66 on the Clearwater River, between RM 108.31 and 138.94 on Lower Granite Reservoir, at RM 139.29 at the confluence of the Snake and Clearwater Rivers, and between RM 139.43 and 141.21 on the Snake River (table 1). The collection of cores was attempted at 86 locations using a piston, gravity, box, or vibrating core sampler.

Collection of Cores

All cores were collected from submerged sediment deposits using a 7.5-meter (m) long pontoon boat with a 4.5-m A-frame (fig. 4). This boat has the advantages of (1) accommodating the use of a hydraulic winch to deploy box, gravity, and piston corers or a vibracorer; (2) providing sufficient height for recovery of long cores (up to 3.5 m); (3) providing plenty of workspace; and (4) providing a stable work environment that is minimally affected by high winds typical of the study area.

Site Selection

Locations for the collection of cores (fig. 1) included cross sections that were measured as part of the 1995 and 2008 bed-sediment surveys done by the USACE (Greg Teasdale, U.S. Army Corps of Engineers, written commun., 2010) (appendix 1). Core sampling locations were grouped closely together (2.9 cross sections per river mile) near the confluence of the Snake and Clearwater Rivers (RM 139.29). To investigate the relative sediment contributions from the Snake and Clearwater Rivers, cores samples were collected in the reach of the Snake River immediately downstream from the confluence of the Snake and Clearwater Rivers and in both rivers upstream from the confluence (fig. 1 inset). Farther downstream from the confluence, extending through the Lower Granite Reservoir to Lower Granite Dam, sampling locations were more widely distributed (10 sections of Lower Granite Reservoir were sampled from RM 135.15 to 108.31).

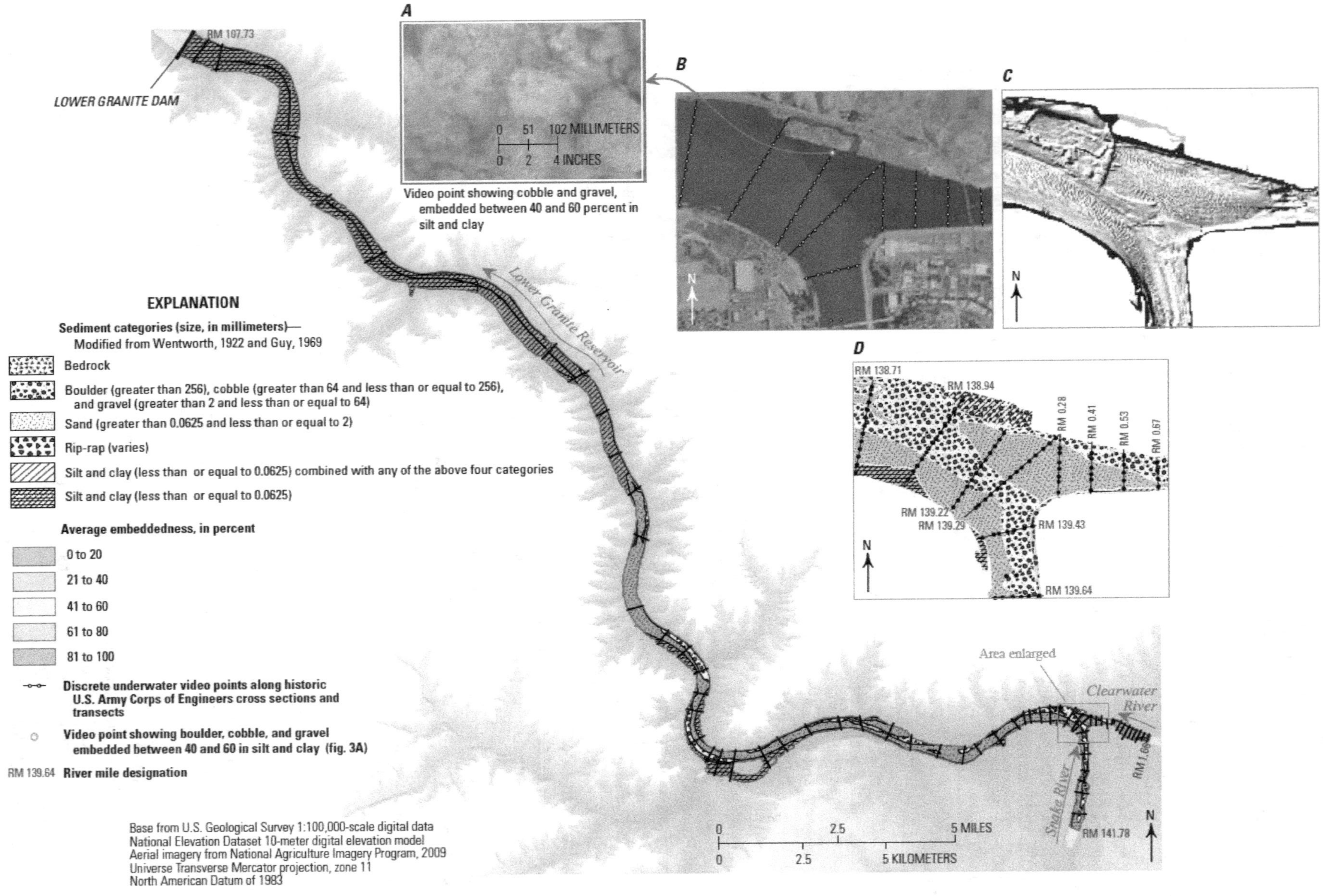
A
B
C
D
RM 107.73
LOWER GRANITE DAM
0 51 102 MILLIMETERS
0 2 4 INCHES
Video point showing cobble and gravel, embedded between 40 and 60 percent in silt and clay
Lower Granite Reservoir
EXPLANATION
Sediment categories (size, in millimeters)—
Modified from Wentworth, 1922 and Guy, 1969
Bedrock
Boulder (greater than 256), cobble (greater than 64 and less than or equal to 256), and gravel (greater than 2 and less than or equal to 64)
Sand (greater than 0.0625 and less than or equal to 2)
Rip-rap (varies)
Silt and clay (less than or equal to 0.0625) combined with any of the above four categories
Silt and clay (less than or equal to 0.0625)
Average embeddedness, in percent
0 to 20
21 to 40
41 to 60
61 to 80
81 to 100
Discrete underwater video points along historic U.S. Army Corps of Engineers cross sections and transects
Video point showing boulder, cobble, and gravel embedded between 40 and 60 in silt and clay (fig. 3A)
RM 139.64 River mile designation
Base from U.S. Geological Survey 1:100,000-scale digital data
National Elevation Dataset 10-meter digital elevation model
Aerial imagery from National Agriculture Imagery Program, 2009
Universe Transverse Mercator projection, zone 11
North American Datum of 1983
0 2.5 5 MILES
0 2.5 5 KILOMETERS
N
RM 138.71
RM 138.94
RM 0.28
RM 0.41
RM 0.53
RM 0.67
RM 139.22
RM 139.29
RM 139.43
RM 139.64
Area enlarged
Clearwater River
Snake River
RM 141.78

Figure 3. Facies of Lower Granite Reservoir and the Snake and Clearwater Rivers in eastern Washington and northern Idaho derived from *A*, video points (to determine sediment size and embeddedness) from the underwater video mapping (UVM) survey recorded at *B*, discrete increments along sections and dredge-material deposit areas used in conjunction with *C*, a bathymetric map created using multibeam echosounding (MBES) survey results, and composite UVM and MBES results shown enlarged as inset *D* (modified from Williams and others, 2012).

Table 1.   Bed-sediment core samples collected from the Lower Granite Reservoir and Snake and Clearwater Rivers in eastern Washington and northern Idaho, 2010.

[NAVD 88, North American Vertical Datum of 1988; m, meters; cm, centimeters; --, no USGS station number established because it was not possible to collect a core sample or the core sample was not submitted for analyses]

USGS station number	Bed-sediment core sample site and core identifier	River mile	Collection date	Latitude (decimal degrees)	Longitude (decimal degrees)	Bottom elevation NAVD 88 (m)	Type of core sampler	Core sample length (cm)	Water depth (m)	Number of grain size subsamples	Number of major and trace element subsamples
				Clearwater River							
462534117015500	1	0.28	5/14/2010	46.4258	117.03242	219	vibracore	21	4.5	2	0
462537117015600	2	0.28	5/13/2010	46.4269	117.03233	220	vibracore	172	4.1	3	0
462535117015700	3	0.28	5/14/2010	46.4265	117.03240	216	vibracore	134	3.8	6	0
462532117014900	4	0.41	5/13/2010	46.4256	117.03026	218	vibracore	52	5.8	2	0
462535117014800	5	0.41	5/14/2010	46.4263	117.03005	219	vibracore	55	4.5	4	0
--	6	0.41	5/11/2010	46.4272	117.03004	217	vibracore and box	0	6.7	0	0
462527117011300	7	0.92	5/18/2010	46.4241	117.02025	220	box	2	4.9	1	0
462532117011101	9	0.92	5/13/2010	46.4254	117.01963	218	vibracore	55	6.4	3	3
462517117010300	10	1.16	5/13/2010	46.4214	117.01745	220	vibracore	110	4.4	4	0
462520117005900	11	1.16	5/18/2010	46.4223	117.01643	222	box[1]	3	2.7	1	0
462524117005900	12	1.16	5/13/2010	46.4233	117.01637	219	vibracore	37	4.6	2	0
462513117004900	13	1.36	5/18/2010	46.4204	117.01368	220	box[1]	4	4.2	1	0
462517117004900	14	1.36	5/18/2010	46.4213	117.01348	221	vibracore	57	3.4	3	0
462520117004700	15	1.36	5/18/2010	46.4223	117.01263	219	vibracore	61	4.2	5	0
462511117002800	16	1.66	5/18/2010	46.4197	117.00767	219	box[1]	3	5.3	1	0
462512117002500	17	1.66	5/18/2010	46.4201	117.00705	219	box[1]	3	5.0	1	0
462515117002400	18	1.66	5/18/2010	46.4208	117.00658	219	box[1]	3	4.4	1	0
				Lower Granite Reservoir							
463914117245400	19	108.31	4/7/2010	46.6538	117.41492	202	piston	213	22.9	3	0
463921117245200	20	108.31	4/7/2010	46.6558	117.41440	191	piston	308	33.5	3	0
463930117244800	21	108.31	4/7/2010	46.6585	117.41330	198	piston	49	26.2	3	0
463749117232500	22	111.24	4/7/2010	46.6304	117.39038	188	piston	172	36.0	3	0
463748117231500	23	111.24	4/7/2010	46.6301	117.38758	191	piston	169	33.5	3	0
463745117225500	24	111.24	4/6/2010	46.6292	117.38183	211	piston	32	11.3	2	0
463516117210100	25	114.92	4/7/2010	46.5877	117.35022	193	piston	49	31.7	3	0
463519117205200	26	114.92	4/7/2010	46.5886	117.34780	196	piston	179	29.0	3	0
463521117204600	27	114.92	4/7/2010	46.5892	117.34603	202	piston	264	21.9	4	0
463316117161800	28	119.56	4/6/2010	46.5543	117.27168	214	gravity	102	12.8	3	0
463318117161600	29	119.56	4/6/2010	46.5551	117.27120	202	gravity	96	21.0	3	0
463322117161000	30B	119.56	4/6/2010	46.5563	117.26935	197	gravity	19	27.1	2	0
463216117150001	31	121.42	4/6/2010	46.5378	117.24992	199	piston	167	25.9	3	13
463219117145900	32	121.42	4/6/2010	46.5387	117.24982	198	piston	91	27.7	3	0
463221117145700	33	121.42	4/6/2010	46.5393	117.24910	200	gravity	27	23.2	2	0

Table 1.   Bed-sediment core samples collected from the Lower Granite Reservoir and Snake and Clearwater Rivers in eastern Washington and northern Idaho, 2010. —Continued

[NAVD 88, North American Vertical Datum of 1988; m, meters; cm, centimeters; --, no USGS station number established because it was not possible to collect a core sample or the core sample was not submitted for analyses]

USGS station number	Bed-sediment core sample site and core identifier	River mile	Collection date	Latitude (decimal degrees)	Longitude (decimal degrees)	Bottom elevation NAVD 88 (m)	Type of core sampler	Core sample length (cm)	Water depth (m)	Number of grain size subsamples	Number of major and trace element subsamples
					Lower Granite Reservoir—Continued						
462655117123600	34	128.27	4/9/2010	46.4485	117.21002	214	piston	266	9.1	5	0
--	35	128.27	4/9/2010	46.4488	117.20862	206	box	2	18.6	0	0
--	36	128.27	4/9/2010	46.4492	117.20662	201	box	3	22.9	0	0
--	37	130.44	4/9/2010	46.4206	117.20688	211	box	4	12.5	0	0
462520117122001	38	130.44	5/12/2010	46.4220	117.20573	210	vibracore	178	14.6	6	6
462520117122001	38A	130.44	4/9/2010	46.4221	117.20558	210	gravity	65	14.3	3	0
462522117121800	39	130.44	5/12/2010	46.4227	117.20501	215	vibracore	164	10.1	4	0
462525117102600	40	132.05	5/12/2010	46.4238	117.1739	209	vibracore	173	12.5	4	0
462529117102800	41	132.05	5/12/2010	46.4247	117.1744	205	vibracore	8	19.1	1	0
462540117081100	43	133.98	5/12/2010	46.4278	117.13626	208	vibracore	140	16.8	4	0
462542117081100	44	133.98	5/12/2010	46.4284	117.13651	205	vibracore	4	19.1	1	0
--	45	133.98	5/12/2010	46.4291	117.13659	206	box	0	16.2	0	0
462514117065400	46	135.15	5/13/2010	46.4205	117.11497	214	vibracore	124	9.8	4	0
462515117065401	47	135.15	5/13/2010	46.4208	117.11497	211	vibracore	124	12.8	3	3
462520117065300	48	135.15	5/12/2010	46.4223	117.11466	211	box	2	12.5	1	0
462456117052900	49	136.29	5/14/2010	46.4156	117.09128	217	vibracore	135	7.5	5	0
462501117052800	50	136.29	5/14/2010	46.4170	117.09121	216	vibracore	151	7.9	3	0
--	51	136.29	5/14/2010	46.4184	117.09071	217	box	0	6.9	0	0
462510117053900	52	136.29	5/14/2010	46.4196	117.09420	215	vibracore	115	10.1	4	0
462506117050400	53	136.69	5/12/2010	46.4182	117.08449	215	vibracore	66	9.6	3	0
462509117050300	54	136.69	5/15/2010	46.4191	117.08418	214	vibracore	123	10.1	3	0
--	55	136.69	5/15/2010	46.4201	117.08514	213	box	0	10.8	0	0
462520117042900	56	137.17	5/15/2010	46.4223	117.07477	217	gravity	20	7.8	1	0
462523117042900	57	137.17	5/15/2010	46.4232	117.07463	213	box	1	12.2	1	0
--	58	137.17	5/15/2010	46.4236	117.07618	210	box	0	14.6	0	0
462540117033500	59	138.07	5/15/2010	46.4277	117.05961	218	box[1]	3	6.7	1	0
462542117033601	60	138.07	5/15/2010	46.4284	117.05988	217	vibracore	70	7.2	3	3
--	61	138.07	5/11/2010	46.4299	117.06026	210	box	0	11.8	0	0
462541117030300	62	138.52	5/12/2010	46.4280	117.05092	220	vibracore	136	3.9	2	0
462546117030500	63	138.52	5/20/2010	46.4294	117.05130	216	piston	147	8.3	2	0
--	64	138.52	5/11/2010	46.4305	117.04965	213	box	0	8.9	0	0
462536117023500	65	138.94	5/11/2010	46.4268	117.04306	220	vibracore	121	4.6	3	0
462542117024801	66	138.94	5/20/2010	46.4282	117.04665	215	piston	76	10.1	2	1
--	67	138.94	5/11/2010	46.4291	117.04117	216	box	0	9.4	0	0

Table 1. Bed-sediment core samples collected from the Lower Granite Reservoir and Snake and Clearwater Rivers in eastern Washington and northern Idaho, 2010. —Continued

[NAVD 88, North American Vertical Datum of 1988; m, meters; cm, centimeters; --, no USGS station number established because it was not possible to collect a core sample or the core sample was not submitted for analyses]

USGS station number	Bed-sediment core sample site and core identifier	River mile	Collection date	Latitude (decimal degrees)	Longitude (decimal degrees)	Bottom elevation NAVD 88 (m)	Type of core sampler	Core sample length (cm)	Water depth (m)	Number of grain size subsamples	Number of major and trace element subsamples
					Confluence						
--	68	139.29	5/19/2010	46.4247	117.03882	216	box	0	8.3	0	0
--	69	139.29	5/11/2010	46.4252	117.03742	215	box	0	9.1	0	0
462535117020701	70	139.29	5/14/2010	46.4263	117.03522	220	vibracore	62	3.8	6	6
462537117020500	71	139.29	5/14/2010	46.4269	117.03468	220	vibracore	162	3.8	5	0
					Snake River						
462522117021500	72	139.43	5/19/2010	46.4228	117.03755	214	box	0.5	10.1	1	0
462522117021501	73	139.43	5/19/2010	46.4228	117.03757	213	box	2	11.3	1	1
--	74	139.43	5/11/2010	46.4230	117.03597	216	box	0	9.3	0	0
--	75	139.64	5/19/2010	46.4192	117.03712	218	box	0	5.9	0	0
--	76	139.64	5/19/2010	46.4191	117.03618	214	vibracore	0	10.5	0	0
--	77	139.64	5/19/2010	46.4192	117.03530	217	box	0	7.9	0	0
462440117021300	78	140.22	5/19/2010	46.4112	117.03693	218	box	0.5	4.4	1	0
--	79	140.22	5/19/2010	46.4106	117.03628	212	box	0	11.4	0	0
462437117020601	80	140.22	5/19/2010	46.4104	117.03505	209	box	1	15.2	1	1
462426117021901	81	140.51	5/19/2010	46.4072	117.03855	215	box	3	9.9	1	1
462426117021600	82	140.51	10/13/2010	46.4072	117.03777	213	box[1]	3	11.9	1	0
462413117022101	84	140.75	10/13/2010	46.4037	117.03927	217	vibracore	110	7.3	5	5
462412117021801	85	140.75	10/13/2010	46.4036	117.03850	218	vibracore	75	7.9	4	4
--	86	140.75	5/19/2010	46.4035	117.03692	218	box	0	6.7	0	0
462348117022801	87A	141.21	5/19/2010	46.3967	117.04110	218	vibracore	19	6.1	1	1
462348117022801	87B	141.21	10/13/2010	46.3965	117.0380	218	vibracore	84	7.3	4	0
462352117022101	88	141.21	5/19/2010	46.3978	117.03917	218	box	3	6.1	1	1
462348117021701	89	141.21	5/19/2010	46.3968	117.03803	217	box[1]	3	7.6	1	1

[1]Box core sampler used after attempts to collect a vibracore or gravity core were unsuccessful.

Three coring sites were sampled at most of the preselected sampling locations (table 1). Coring sites were positioned along each section of river or reservoir where samples were collected to maximize the following selection criteria: thickness of lacustrine sediment (based on bathymetric survey data), sampling of differing depositional environments, and spacing between coring locations. One of the coring locations typically was selected at or near the deepest point within the channel along each section sampled.

Tool Selection

The advantages and disadvantages of each coring tool were taken into account when determining the most appropriate tool for a given sample. One consideration is the thickness of sediment. In reservoirs with thick sediment deposits, such as Lower Granite Reservoir, gravity and piston corers as well as a vibracorer are used because they are able to collect a much longer core compared to a box corer. Gravity

Figure 4.   Pontoon boat with A-frame at Lower Granite Reservoir, 2010.

and piston corers are capable of collecting a 3-m core that is 6.7 cm in diameter, whereas the vibracorer is capable of collecting a 3.2-m core that is 7.6 cm in diameter. The box corer, which is 14 (length) × 14 (width) × 20 (height) cm, provides a shorter core but collects more material for a given interval compared to the tubular gravity and piston corers, and also has the advantage of collecting a less-disturbed sample (figs. 5A and 5B). Box corers were used predominantly for reconnaissance purposes (particularly at sites that were not expected to have thick sediment deposits) because of the relative ease of use compared to the tubular samplers. However, in some cases, box cores were subsampled for grain size and major and trace elements on the boat at the coring location.

The choice between using a gravity or piston corer is influenced by multiple factors. Gravity corers likely collect a less disturbed core and are easier to use. As a result, gravity corers were used often, especially for reconnaissance purposes. However, gravity corers have one important limitation—core shortening. Core shortening refers to the thinning of sediment layers recovered relative to undisturbed sediment; it has been attributed to friction on the walls of the liner (Emery and Hulsemann, 1964). In addition to friction, the sediment in the gravity corer barrel must push water out the top through a check valve, which generates back pressure inside the liner. In 1998, a piston and a gravity core were collected side by side in White Rock Lake, Tex. (Van Metre and others, 2004). Both encountered prereservoir sediment and appeared to represent the complete sediment sequence on the basis of color banding in the cores. The thickness of lacustrine sediment in the gravity core was 122 cm compared to 206 cm in the piston core, a core shortening of 41 percent. Therefore, if quantifying total lacustrine thickness is a primary objective, as it was in this study, an alternative coring method to gravity coring should be used, or an accounting of core shortening should be done (Juracek, 1998).

Vibrating core samplers (vibracorers) have an advantage over coring devices that sink (box corers), drop (gravity corers), or propel (piston corers) (Van Metre and others, 2004); unlike these other samplers, the vibracorer generates a high frequency vibration that transfers more energy to the sediment—greatly reducing wall friction inside and outside of the tube used to collect the core. This results in longer, more representative cores. One disadvantage of vibracoring occurs when wall friction inside the tube (as the tube penetrates sediment) exceeds the bearing strength of the sediment causing the sediment inside the tube to stop moving even though the tube continues to penetrate the sediment. This can lead to intermediate layers of sediment being bypassed unbeknownst to the investigator; this effect is referred to as plugging or rodding. Additional drawbacks associated with vibracoring include the potential for resuspension of the top few centimeters of water-rich sediment and possible compaction of sand and organic-rich layers by the associated vibrations (Vibracoring Concepts, 2011).

Coring

One or more cores were collected at selected core sampling sites for grain size distribution analysis and for trace element analysis following procedures described by Shelton (1994). In most cases, only one core was collected unless the sediment thickness was less than expected based on the river-bed elevation and geomorphology from the bathymetric survey. In these instances, a second or even third core would often be collected, frequently using an alternative coring method. Latitude and longitude of the coring site were obtained from a global positioning system to ensure that the location being sampled was at or near the predetermined sampling location. The depth of water at each site was obtained using a fathometer, and the approximate thickness of sediment obtained in each core was measured (fig. 5C) and recorded.

The gravity corers used in this study had a steel barrel with a polybuterate liner or an aluminum core barrel with lead weight attached and a polybuterate liner held in place by a polyvinyl chloride flange and hose clamps (fig. 5D). Each liner had a check-valve attached to the top, which allowed water to escape during penetration and then closed to help retain sediment during recovery. A cutting head was attached to the bottom of the steel barrel to help protect the liner and penetrate firmer sediments. A core catcher was inserted at the bottom of the liner to help retain sediment during recovery. The corer typically was allowed to free fall during sample collection to maximize sediment recovery. During retrieval, the bottom of the core was immediately capped as it approached the water surface. The cap was taped on while the core was suspended vertically from the A-frame, and the core was removed from the barrel and stored upright until subsampling. Core liners were cut about 1 cm above the top of the sediment to drain any water overlying the sediment (fig. 5E) and then capped and taped to minimize disturbance during transport to the subsampling location. A small gravity corer with an aluminum core barrel (fig. 5D) was used as a reconnaissance tool to estimate sediment thickness and determine the predominant sediment type. Reconnaissance was particularly important in areas where little sediment thickness was expected.

The piston corer works like a syringe with the bottom cut off to create an open cylinder—the piston acts as the plunger and the core barrel acts as the outside of the syringe. The plunger (piston) is held in place just above the sediment and the outside of the syringe is pushed past the piston into the sediment. The piston corer used in this study was the same weight with the same barrel as the gravity corer. However, it contained a piston inside the liner connected to a trigger arm located above the corer and attached to the winch on the boat (figs. 6A and 6B). When the trigger weight suspended from the trigger arm reaches the bottom, the arm releases the corer allowing it to fall past the piston into the sediment. The piston should stop just above the top of the sediment by the cable

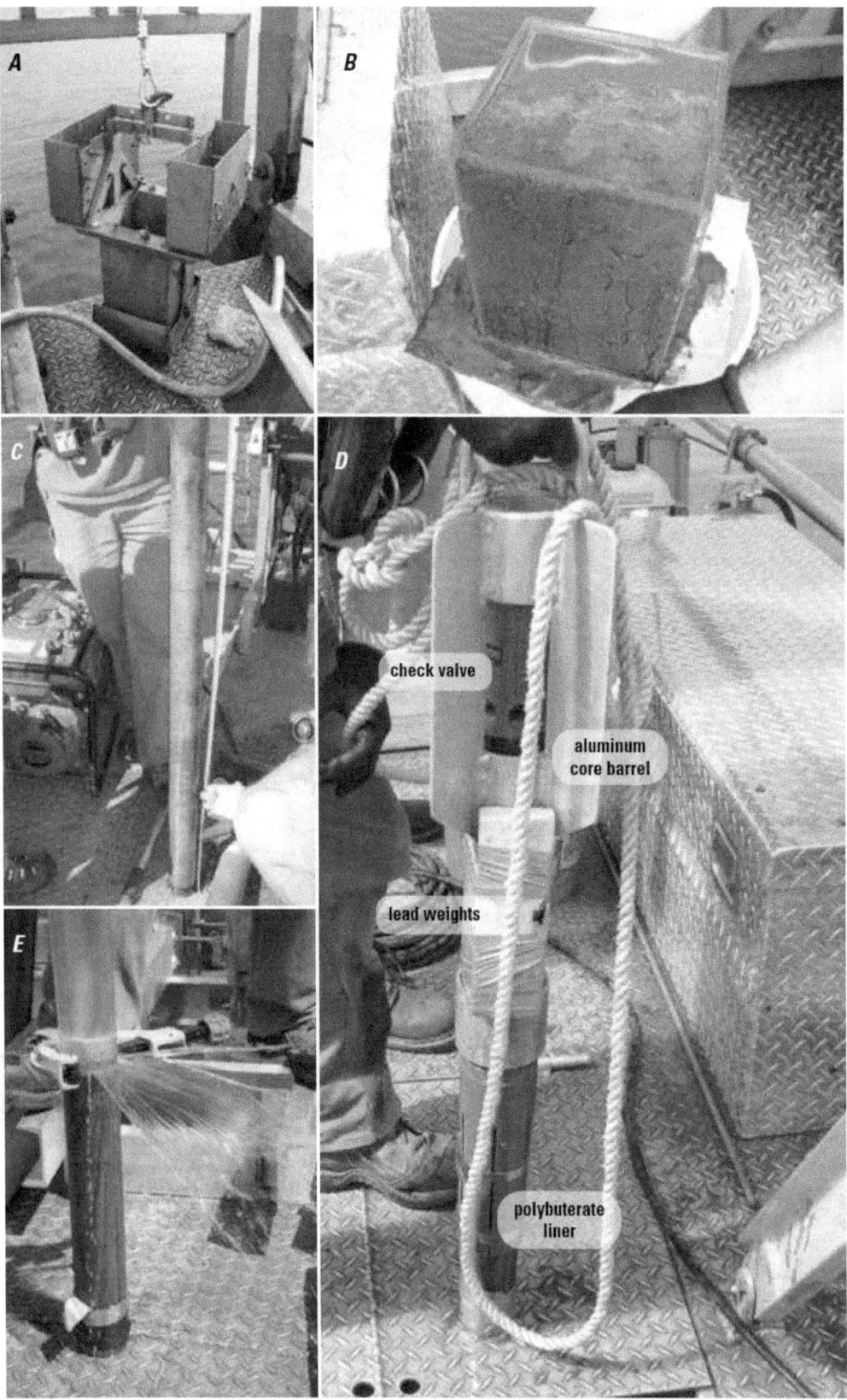

Figure 5.   Different bed-sediment coring tools and procedures, including: *A*, a box corer, *B*, a box core, *C*, measurement of sediment-core thickness, *D*, components of a short aluminum gravity corer, and *E*, draining water overlying a bed-sediment core by cutting the liner with a tubing cutter.

Figure 6.   U.S. Geological Survey personnel *A*, lowering piston corer into the water and *B*, preparing to remove a bed-sediment core from the steel core barrel.

attached to the winch if the length of cable between the trigger arm and piston and the length of rope between the trigger arm and trigger weight were properly measured and the winch operator stopped lowering as soon as the trigger arm released the corer. As the barrel falls past the piston into the sediment, a strong vacuum is created below the piston, which enhances the recovery of sediment. The winch pulls the piston to the top of the corer, if it was not already there, and lifts the corer. Once out of water, the trigger arm and weight are disconnected from the cable. The methods associated with piston core retrieval and storage were comparable to those associated with gravity core retrieval and storage.

The vibracorer was used to collect cores from substrates with predominantly sand or larger particle sizes. The vibrating mechanism of the vibracorer, referred to as the vibrahead (fig. 7A), is powered by an external electrical source and generates 3,000 to 11,000 vibrations per minute. These vibrations cause a thin layer of material to mobilize along the inner and outer walls of the core barrel or liner, reducing friction and easing penetration into the substrate. Metal tubes conduct vibration energy best, with hard steel performing better than aluminum. Plastic tubes are poorer conductors than metals, but this disadvantage is partially offset by the reduced mass requiring vibration (Vibracoring

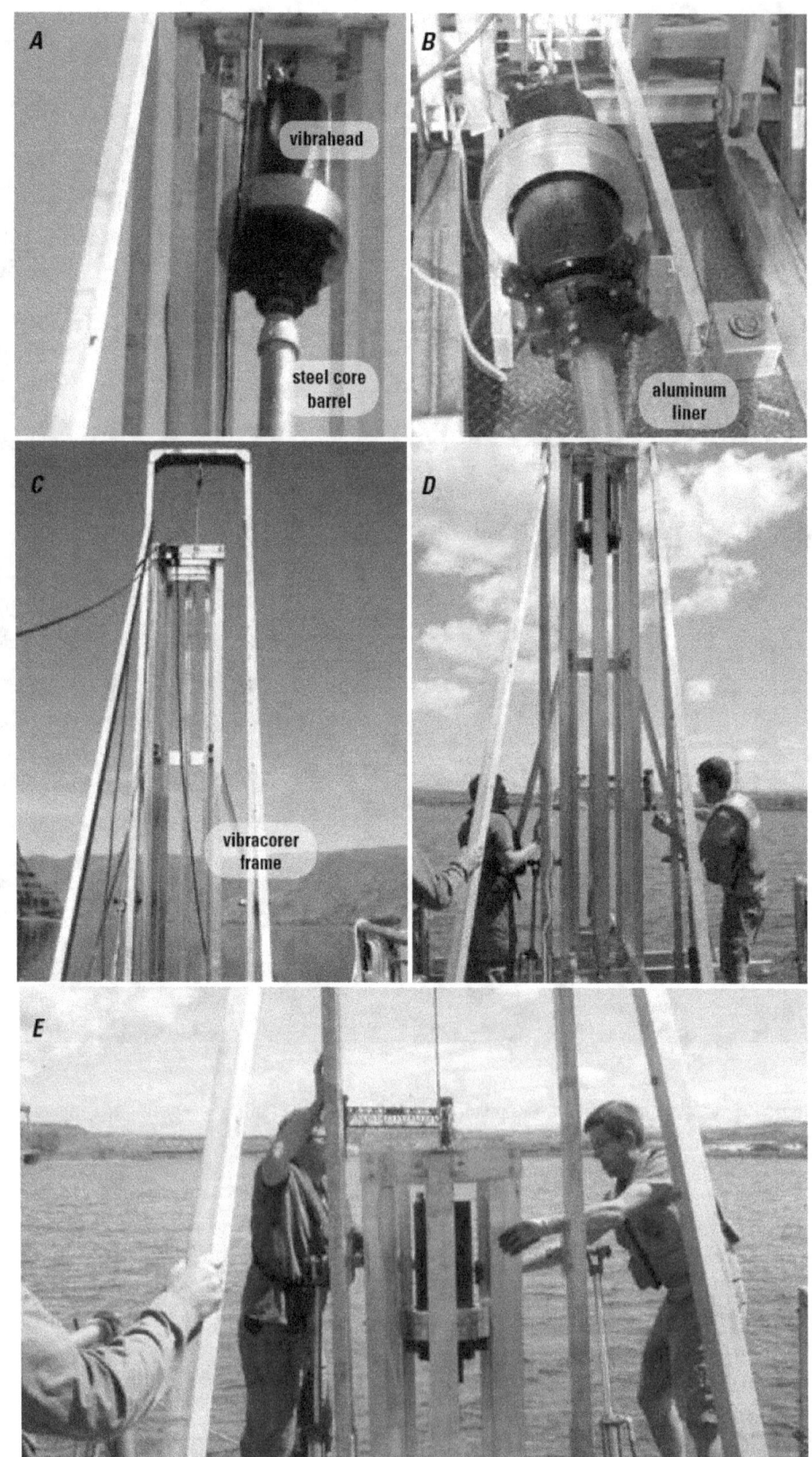

Figure 7.   Use of vibracoring system *A*, use of vibracorer with steel core barrel and polybuterate liner, *B*, use of vibracorer with aluminum liner and no barrel, *C*, vibracoring frame resting on the deck of the boat, *D*, preparation for deployment of vibracorer, and *E*, lowering of vibracorer inside frame.

Concepts, 2011). For this study, the same steel barrels and polybuterate liners used to collect the gravity and piston cores were used initially to collect the vibracores (fig. 7A). However, the vibracorer did not appear to be penetrating the sediment to the expected depth. This was likely due to a loss in vibration efficiency because of the small gap between the liner and the barrel. Therefore, a transition was made to using aluminum liners without a core barrel (fig. 7B). In order to keep the vibracorer from penetrating the sediment at an angle, a frame was constructed to house the vibracorer and ensure penetration perpendicular to the sediment surface (figs. 7A, 7B, and 7C (Vibracoring Concepts, 2011). The vibracorer frame was lowered from the front end of the pontoon boat using an onboard winch until the base of the frame was resting on the bottom sediment of the reservoir or river being sampled (figs. 7D and 7E). Once a vibracore was collected (using an aluminum liner), a handsaw was used to remove most of the unused liner above the uppermost sediment layer in the core; any water that was collected above the top of the sediment in the core also escaped at this time. Because the aluminum core liner is opaque, locating the top of the sediment core was problematic. Therefore, using a handsaw, the liner was cut several centimeters above where the top of the sediment core was anticipated. The last few centimeters of liner above the top of the core were then removed using a tubing cutter (fig. 5E).

Subsampling and Description of Cores

Piston cores, gravity cores, and vibracores were held vertically during removal from the samplers and transported to the river bank or reservoir shore, where they were split lengthwise, photographed, described, and subsampled. Core extraction from bottom sediments is referred to as sampling, whereas the subdivision of each core into discrete intervals is referred to as subsampling. The cores were laid on the ground and split lengthwise using a circular saw to cut through most of the liner, cutting the rest of the liner with a utility knife, and then slicing the sediment with a stainless steel or Teflon spatula. The core was then split open next to a tape measure, photographed, and described. Descriptions included color, texture, odor, and the presence or absence of organic detritus and biota or other visible debris. Surficial sediment samples were designated as any sample that included some part of the top 5 cm of the core. Identification of the prereservoir land surface (if penetrated) was an important part of the description. Lacustrine sediments typically have a high water content (porosity of 70 to 90 percent) and are usually silt and clay, whereas prereservoir soils are drier or stickier compared to the overlying lacustrine sediment, and often have root hairs and gravel (sediment larger than 2.0 mm in diameter). The top of the prereservoir surface may also have a layer of decaying leaves and sticks.

After a core was described and photographed, one-half of the core was cut into sections and placed into a plastic

core archival box. The sediment depths of each core section were noted on the core archival box. Archived cores were provided to the USACE in Walla Walla, Wash., for long-term storage. Samples for laboratory analyses were collected from the remaining one-half of the core. Samples were collected from each section of the core with a distinct grain size. For example, the sediment in core sample 38 collected on May 12, 2010, changed at a depth of 101.4 cm from a gray colored, medium to fine–sized sand (0.25 to 0.125 mm in diameter) to dark brown silt and clay (fig. 8). Two to five samples were collected from most gravity cores, piston cores, and vibracores for grain-size analysis. One sample was collected from most box cores for grain-size analysis. The samples were collected by scooping the sediment out of the core liner at a selected depth interval with a Teflon spatula and transferring it to a sample container. Samples for grain-size analysis were placed in precleaned polypropylene jars. Samples for analysis of major and trace elements were placed in separate precleaned polypropylene jars. Each sample was labeled with the river mile, core identification, and depth interval of the core, in centimeters. Sampling tools were rinsed in tap water, soaked and washed with a brush in phosphate-free detergent, and rinsed again in tap water between samples.

Analytical Methods

Sediment samples were analyzed for grain size and selected major and trace elements. All samples selected for grain-size analysis were sent to the USGS Cascades Volcano Observatory Laboratory in Vancouver, Wash. Samples for major and trace element analyses were sent to the USGS Mineral Resources Program Analytical Laboratories in Denver, Colorado.

Grain Size

Two methods were used for grain-size analysis. For coarse material (sand-sized particle larger than 0.0625 mm in diameter), the wet sieve method was used (Pope and Ward, 1998). For silt and clay sized particles, a Micromeritics SediGraph 5120 was used to perform a sedimentation technique (referred to hereinafter as the Meyer-Fisher sedimentation technique), which measures the gravity-induced settling rates of different size particles in a liquid with known properties as a means to provide grain-size information (Meyer and Fisher, 1997). Sand-fine separation and scale of size classes defined in Wentworth (1922) and Guy (1969) for gravel, sand, silt, and clay particles were followed.

A sand-fine separation was initially done to determine which grain-size classification method would be used. If less than 5 percent of the sample by weight was sand, then a total sand weight was reported, and the Meyer-Fisher sedimentation technique was used to analyze the silt and clay (fines). If the sample had less than 5 percent silt and clay by weight, then a total weight for the fines was reported, and the wet sieve

Figure 8. Bed-sediment core 38 collected at river mile (RM) 130.44 showing two sections of the core with distinct grain sizes for core intervals above and below a depth of 101.4 centimeters (cm) and intervals where grain-size samples were collected.

method was used to analyze the coarse sediment. If the sample had enough material from both size classes, then a complete size analysis was performed. Because it was necessary to keep the material wet throughout the analysis, the decision for a complete size analysis was based on a visual determination by the technician performing the analysis. If the amount of silt and clay material was less than 5 percent, then generally there was not enough material to perform the Meyer-Fisher sedimentation technique (appendix 2).

Major and Trace Elements

Freeze-dried sediment samples were analyzed for major and trace elements by the USGS Mineral Resources Program Analytical Laboratories in Denver, Colo., for the USGS National Water Quality Laboratory. Samples for major and trace element analyses were digested completely using a mixture of hydrochloric, nitric, perchloric, and hydrofluoric acids and analyzed by inductively coupled plasma–mass spectrometry (ICP–MS) (Briggs and Meier, 2002). Concentrations of mercury were determined by continuous flow-cold vapor-atomic fluorescence spectrometry (CVAFS) (Hageman, 2007) (appendix 3).

Quality Assurance and Quality Control

Quality-control (QC) samples for the major and trace elements were analyzed to ensure the quality, precision, accuracy, and completeness of the dataset; no QC samples were analyzed for grain size. The QC samples for major and trace elements consisted of laboratory reagent blank samples, standard reference materials (SRMs), and replicate samples analyzed with each set of environmental samples (appendix 4). The Lower Granite Reservoir samples were analyzed for major and trace elements in two different sets. A "lower reporting limit" described by Taggart (2002, p. viii) as greater than or equal to "five times the standard deviation determined from the method blank" was used for major and trace elements. Major and trace element concentrations less than the respective lower reporting limits are hereinafter referred to as nondetections, and concentrations equal to or greater than the respective lower reporting limits are hereinafter referred to as detections.

At least one major or trace element was detected in 13 percent of the blank analyses; cesium, niobium, and antimony were the most frequently detected trace elements. There were no detections in the blank analyses for 21 of the 38 elements. The laboratory analyzed SRMs, compared the values to published or standard results, computed the percent recoveries, and noted unacceptable SRM recoveries that were 10 percent more or less than 100 percent recovery. The SRM results are used to assess bias for each analysis. Titanium recoveries were less than 90 percent in 13 of 16 SRM analyses, whereas gallium recoveries were greater than 110 percent in 11 of 16 SRM analyses. Other elements with 50 percent or more unacceptable SRM recoveries included bismuth, cerium, and yttrium.

The laboratory analyzed two laboratory replicates in which the sample was split at the laboratory (appendix 4). Replicates provide information about variability in the analytical process, but results can be affected by sample heterogeneity, particularly when sediments are the sample media (Pirkey and Glodt, 1998). The relative percent difference (RPD) was computed between each pair of replicate analyses to provide a measure of precision using the equation:

$$RPD = |C_1 - C_2|/((C_1 + C_2)/2) \; x \; 100, \qquad (1)$$

where

C_1 is the constituent concentration, in micrograms per gram, of one observed value; and

C_2 is the constituent concentration, in micrograms per gram, of a second observed value.

The overall average RPD for the major and trace element replicate analyses was 3.7 percent. The highest RPDs were for replicate analyses of antimony (88.8, 84.7, and 33.8 percent), followed by silver (29.4 percent).

Grain-Size Distribution

Results from the three data collection techniques used in the multiyear study (MBES bathymetric survey map, facies map based on UVM survey results, and grain-size analyses of sediment cores) were consolidated to better understand the grain-size distribution. This data consolidation also provided information on variations in substrate type relative to embeddedness throughout the reach as well as the predominance of fines (silt and clay) in the backwater-affected area downstream from the confluence. It also provides an opportunity to interpret surficial sedimentary structures (bedforms) of Lower Granite Reservoir, the Snake and Clearwater Rivers, and the confluence of these two rivers (fig. 9).

Results from the grain-size analyses of sediment cores also provided a quantitative mechanism for verifying the facies map generated from the UVM surveys (Williams and others, 2012) (fig. 10). None of the 25 surficial sediment samples with less than 20.1 percent silt and clay (based

on grain-size analysis) were located in areas that the facies map identified as having silt and clay size particles. Of the 12 surficial sediment samples with greater than 80 percent silt and clay (based on grain-size analysis), 10 were located in areas that were identified as having silt and clay size particles based on the facies map; the two remaining samples were located in areas classified as sand on the facies map.

Percent silt and clay in surficial sediment samples obtained from bed-sediment cores collected in the Snake and Clearwater Rivers and Lower Granite Reservoir are shown in figure 10. Sites at the lower end of the reservoir, closest to the dam (sites 19–30) where stream velocities are lower, tended to have larger volumes of silt and clay compared to the other sites that were surveyed. All of the surficial sediment samples collected in the Snake River upstream from the confluence had less than 20 percent silt and clay. This is most likely because velocities in this reach of the Snake River are high enough to keep fine-grained (silt and clay) sediment particles entrained. Most of the surficial sediment samples collected in the Clearwater River (9 out of 13) contained less than 40 percent silt and clay; only one site (site 16) had more than 60 percent silt and clay, and this site was located in a near-bank, lower-velocity margin environment. Surficial sediment samples collected near midchannel at the confluence (site 70) tended to have more silt and clay than most surficial sediment sample collection sites on the Snake and Clearwater Rivers or even sites further downstream in Lower Granite Reservoir (sites 59, 60, and 63). The turbulence and reduction in velocity induced by the confluence of the Snake and Clearwater Rivers likely caused these two rivers to drop much of their sediment load in this area. Of the remaining surficial sediment samples (those located downstream from site 70 and upstream from site 30), all but two (site 52, which is located in a low-velocity, near-shore environment and site 66, which is located on the inside of a bend in Lower Granite Reservoir) had less than 40 percent silt and clay.

Downcore grain-size data from cores collected near the confluence of the Snake and Clearwater Rivers were plotted (fig. 11) to explore grain-size variations in this transitional area from free flowing to backwater conditions. Of the four proposed coring locations at the confluence, cores could only be collected at two locations, sites 70 and 71. Attempts at collecting box cores at the two remaining confluence sites, 68 and 69, resulted in nothing more than a trace amount of sand at site 69. Site 68 was located in an area identified as sand on the facies map and had visible dunes in the images generated from the MBES survey (Williams and others, 2012), so the inability to collect a core at this location was unexpected; rivers are dynamic systems and locations of dunes can change gradually over time or rapidly in response to storm events (Germanoski and Schumm, 1993). Site 69 was located in an area defined as boulder, cobble, and gravel on the facies map, so the inability to collect a core at this location was expected.

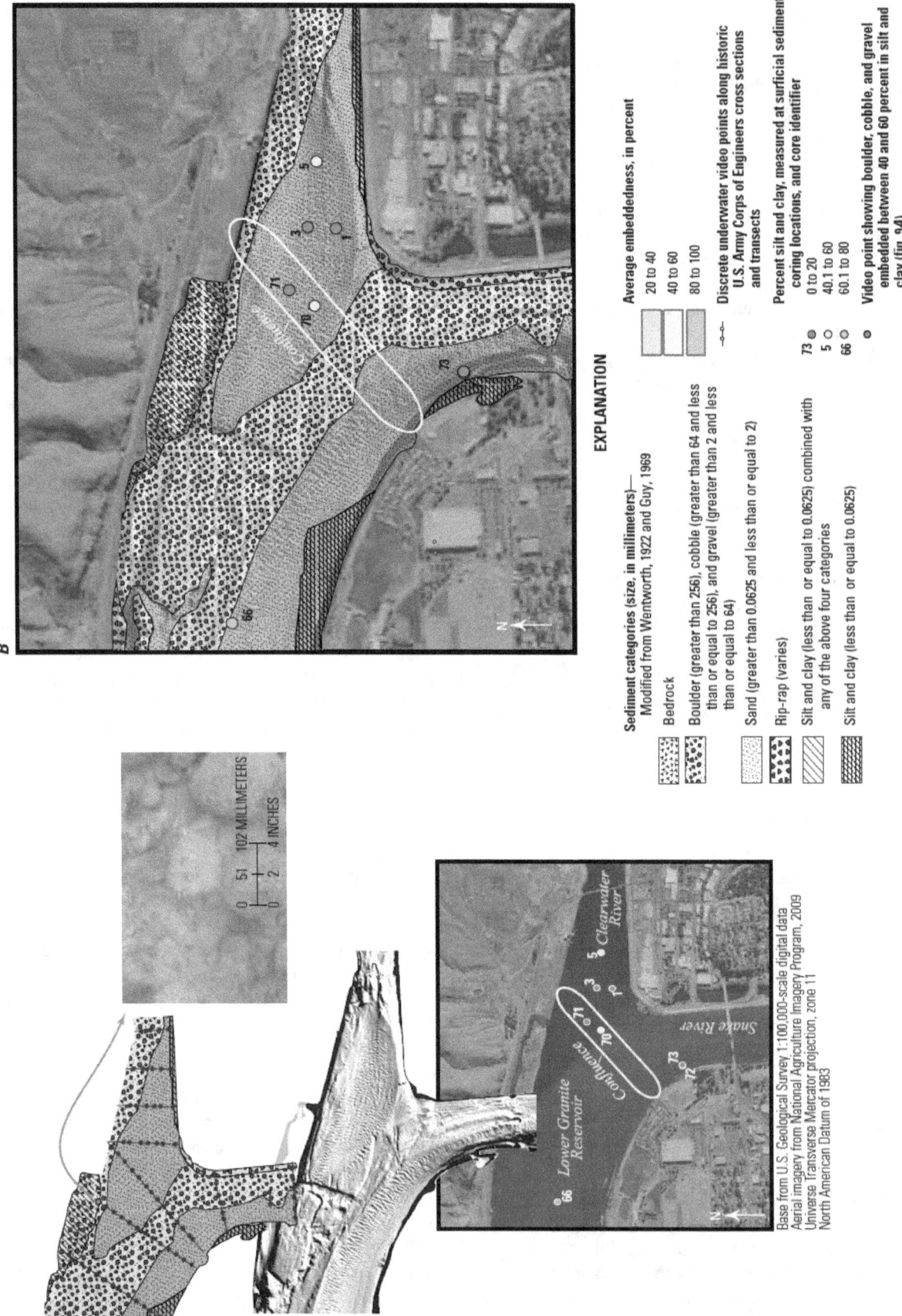

Figure 9. *A,* Consolidation of data from various data-collection methods, including: facies map (derived from underwater video map survey results), bathymetry (derived from multibeam echosounding survey results), and grain-size data from surficial sediment coring and laser projection to generate *B,* a composite map of these data layers.

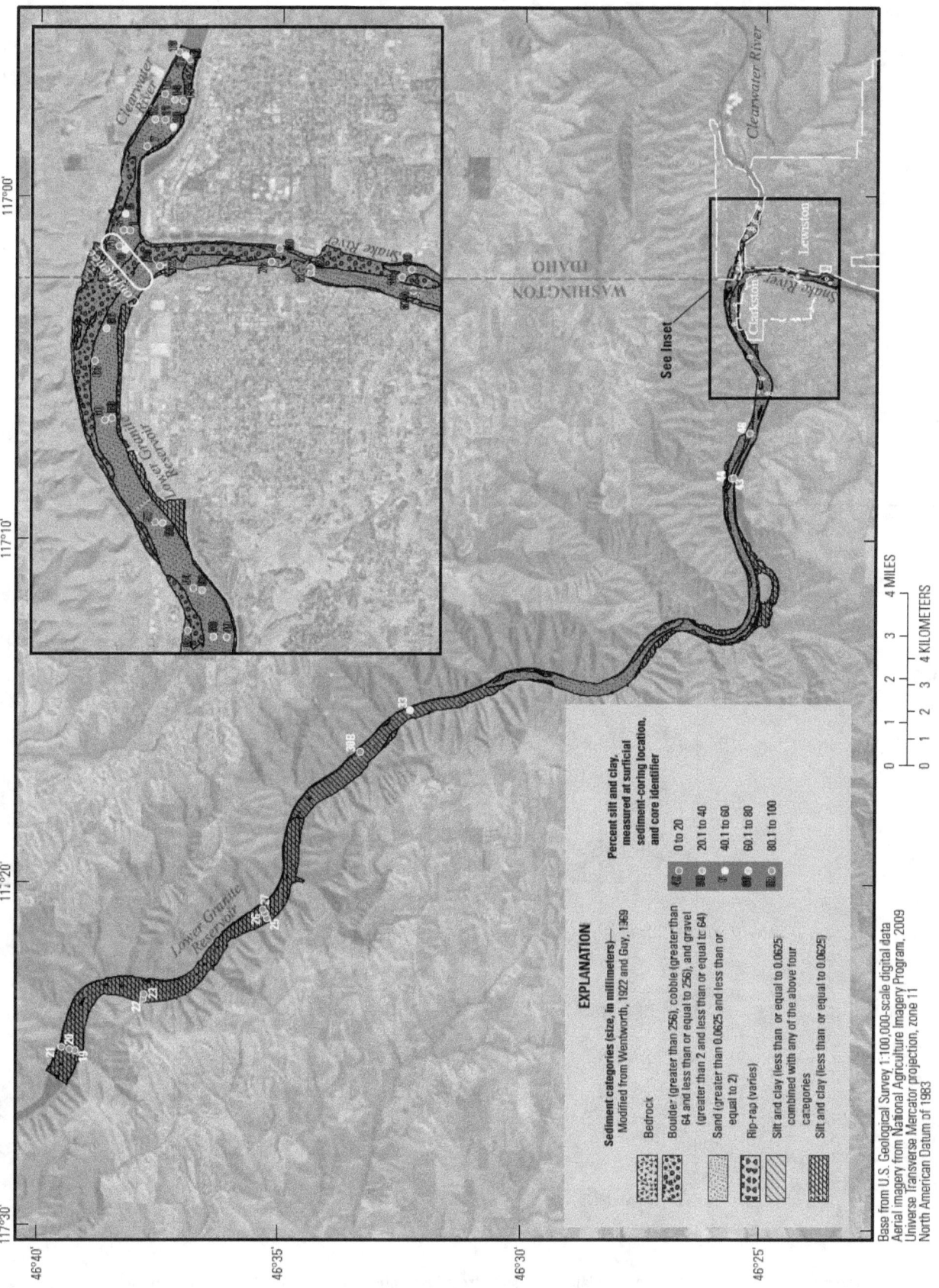

Base from U.S. Geological Survey 1:100,000-scale digital data
Aerial imagery from National Agriculture Imagery Program, 2009
Universe Transverse Mercator projection, zone 11
North American Datum of 1983

Figure 10.   Percent silt and clay in surficial-sediment samples obtained from bed-sediment cores collected in Lower Granite Reservoir and the Snake and Clearwater Rivers, eastern Washington and northern Idaho, 2010.

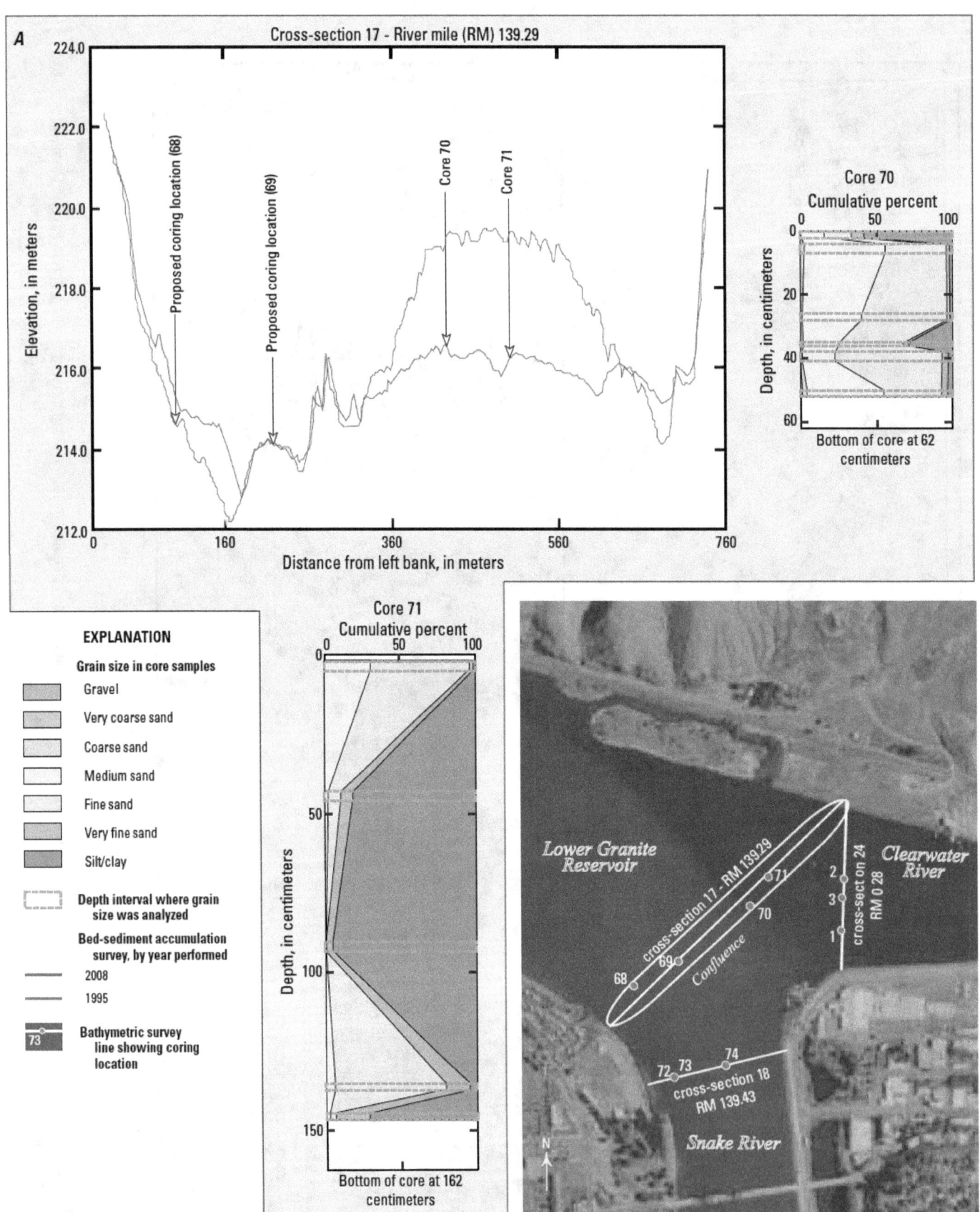

Figure 11. Comparison of grain-size distribution (as it relates to results from bed-sediment accumulation surveys done by the U.S. Army Corps of Engineers in 1995 and 2008) in bed-sediment cores collected near the confluence of the Snake and Clearwater Rivers at *A,* cross-section 17, *B,* cross-section 18, and *C,* cross-section 24, eastern Washington and northern Idaho, 2010.

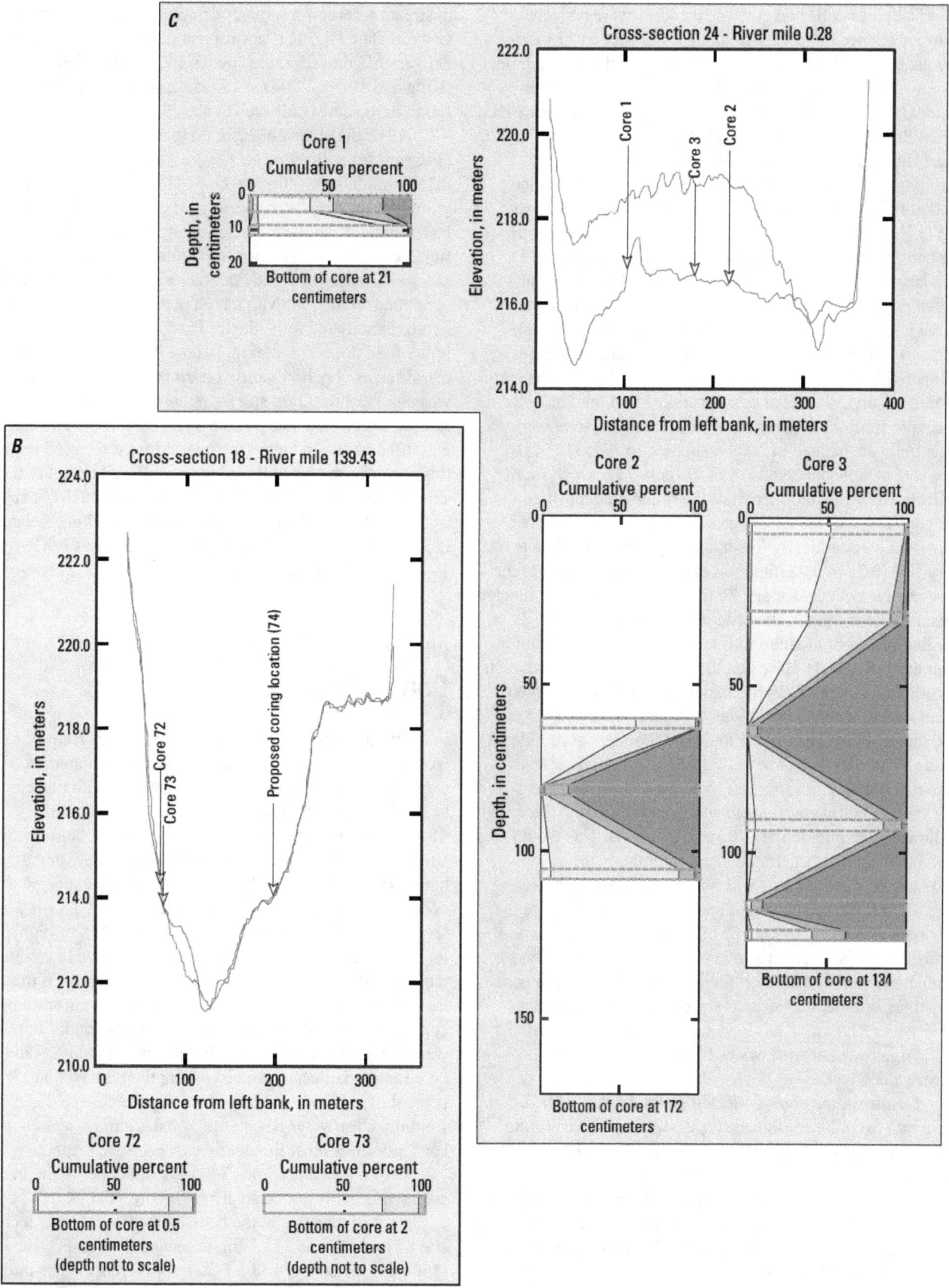

Figure 11. Comparison of grain-size distribution (as it relates to results from bed-sediment accumulation surveys done by the U.S. Army Corps of Engineers in 1995 and 2008) in bed-sediment cores collected near the confluence of the Snake and Clearwater Rivers at *A*, cross-section 17, *B*, cross-section 18, and *C*, cross-section 24, eastern Washington and northern Idaho, 2010.—Continued

The cores collected at sites 70 and 71 were collected from a thick sediment deposit along cross-section 17 just downstream from the confluence. Following the bed-sediment survey in 1995, floods with peak discharges of more than 80,000 ft³/s occurred in both 1996 and 1997 on the Clearwater River. Streamflow records from U.S. Geological Survey streamflow-gaging station 13342500 Clearwater River at Spalding, Idaho, document peak streamflows of more than 85,000 ft³/s during 1996 and 1997 (U.S. Geological Survey, 2011); this gaging station is about 16 miles upstream from the confluence of the Snake and Clearwater Rivers (fig. 1). The large peak streamflows likely deposited large amounts of sediment and substantially changed the channel geometry of the Clearwater River; 2008 channel-geometry surveys at cross-sections 17 and 24 document appreciable changes compared to 1995 (figs. 11A and 11C, respectively). Channel-geometry surveys done at cross-section 18 on the Snake River upstream from the confluence with Clearwater River were essentially unchanged in 2008 compared to 1995 (fig. 11B). The core sample collected at site 71 (core 71), which was collected near the middle part of the sediment deposit at the confluence of the Clearwater River and Snake Rivers, consisted predominantly of silt and clay with some fine to very fine (0.25 to 0.0625 mm in diameter) sand, whereas the core sample collected at site 70 (core 70), which was collected closer to the margin of the sediment deposit, consisted predominantly of medium to fine sand (0.50 to 0.125 mm in diameter) (fig. 11A). Based on the results of the bed-sediment accumulation survey, approximately 2.6 m of bed sediment accumulated where core 70 was collected between 1995 and 2008, and approximately 3.1 m of bed sediment accumulated at site 71 during that same period. Core shortening effects associated with vibracorer use likely resulted in sediment-core thicknesses that were approximately one-half of the measured sediment accumulation at these two locations. This would mean that core 71 penetrated to approximately the same bed-surface elevation that was observed in 1995, whereas core 70 only penetrated about one-half of the distance to the 1995 bed surface. This may be because of the more coarse-grained material encountered in core 70 (very coarse sand from 1 to 2 mm in diameter) towards the bottom of the core. The thick sediment deposits at the confluence and on the Clearwater River may be flood-related deposits associated with large peak streamflows in 1996 and 1997 on the Clearwater River.

Neither of the cores collected on the Snake River just upstream from the confluence (cross-section 18, river mile 139.43) contained silt and clay (fig. 11B). Both cores were very short (the core sample collected at site 72 was 0.5 cm long; the core sample collected at site 73 was 2 cm long) and made up predominantly of medium sand. An attempt also was made to collect a box core at site 74, but no core could be collected at this location because of minimal sediment accumulation at this location since the 1995 bathymetric survey (fig. 11B). The bottom material in the Snake River upstream from the confluence, in general, is composed predominantly of gravel, cobble, and boulders as a result of

upstream reservoir trapping of fine-grained sediment by Hells Canyon Dam (fig. 1) (Parkinson and others, 2003) and the delivery of coarser grained materials from the Salmon River (King and others, 2004). As a result, scant silt and clay were found in the cores collected at sites 72 and 73.

All of the cores collected on the Clearwater River just upstream from the confluence (cross-section 24, river mile 0.28) contain some silt- and clay-sized particles (fig. 11C). No grain-size analyses were performed on the top 60 cm of the bed-sediment sample collected at site 2 (core 2), but based on field notes, this interval was predominantly fine to medium sand with lenses of woody debris and clay. Cores 2 and 3 contained an interval with a high percentage of silt and clay at approximately the same depth. Based on the core descriptions in the field notes, an interval in core 2 from 76 to 99 cm was classified as clay (the interval from 80 to 83 cm in this core was analyzed for grain size). An interval in the bed-sediment sample collected at site 3 (core 3) from 56 to 87 cm was classified as silt and clay (the interval from 62 to 65 cm in this core was analyzed for grain size). Based on bathymetric survey images (fig. 2) (Williams and others, 2012), it appears as though these silt and clay deposits may be flood-related deposits associated with large peak streamflows in 1996 and 1997 on the Clearwater River.

Major and Trace Element Concentrations

Of the 69 cores collected, 50 subsamples from 15 cores were analyzed for major and trace elements (appendix 3). Concentrations of trace elements were low, with respect to sediment-quality guidelines (SQGs), in most cores. There are typically two SQGs—a lower level, below which adverse effects to aquatic biota are not expected, and a higher level, above which adverse effects are expected to occur. The threshold effect concentration (TEC), or lower level, and probable effect concentration (PEC), or higher level (MacDonald and others, 2000), were used to evaluate the concentrations of trace elements in the cores. Of the trace elements with TECs (arsenic, cadmium, chromium, copper, mercury, nickel, lead, and zinc), concentrations were greater than TECs in 43 percent of the analyses in the core sample collected at site 31 from the reservoir and 29 percent of the analyses in the core sample collected at site 9 from the Clearwater River (fig. 12). Chromium and copper concentrations most frequently exceeded the respective TECs. Two concentrations, 740 µg/g of copper in the bed-sediment sample collected at site 9 (core 9) at 34 to 37 cm and 2.32 µg/g of mercury in the bed-sediment sample collected at site 81 (core 81) at 0 to 3 cm, exceeded their respective PECs (150 µg/g for copper and 1.1 µg/g for mercury). The outlier trace element concentrations that were greater than the PECs were verified by the analytical laboratory (LaDonna Choate, U.S. Geological Survey, National Water Quality Laboratory, written commun., May 10, 2011).

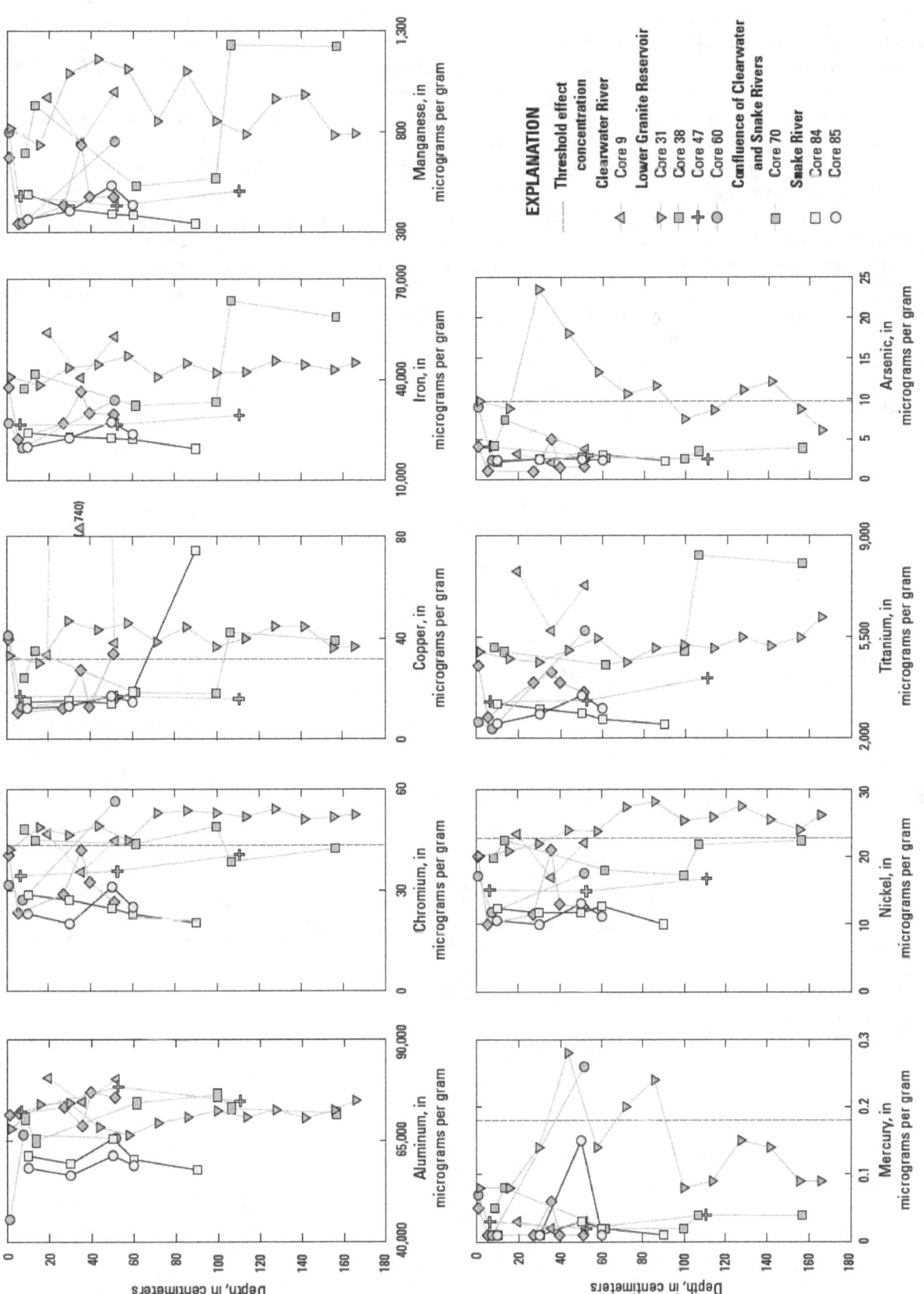

Figure 12. Concentrations of selected major and trace elements with depth (where a depth of 0 centimeters is the top of the core) in selected bed-sediment cores collected in Lower Granite Reservoir, the Snake and Clearwater Rivers, and the confluence of the Snake and Clearwater Rivers, eastern Washington and northern Idaho, 2010.

Major and trace element concentrations varied substantially at different coring locations and with depth in a single core (fig. 12). Typically, concentrations were lower in the samples collected from the Snake River compared to those collected from the Clearwater River, the confluence of the Snake and Clearwater Rivers, or Lower Granite Reservoir. Large variations in grain size occurred with depth in the sediment cores. The percentage of silt and clay-sized particles varied from 0.2 to 96.1 percent in the cores that were analyzed for major and trace elements. Because of the large grain-size variability, it is not possible to describe patterns in major and trace element concentrations with depth in the cores.

Major elements in sediment generally indicate the minerals present. For example, high aluminum and iron concentrations are typically associated with clay minerals that, because of their finer grain size, can preferentially sorb trace elements (Horowitz and Elrick, 1987). All of the samples in the cores, except core 31, that were analyzed for major and trace elements were also analyzed for grain size. Generally, lower concentrations of major and trace elements were associated with coarser sediments (larger than 0.0625 mm in diameter) and higher concentrations of major and trace elements were associated with finer sediments (smaller than 0.0625 mm in diameter) (fig. 13). Samples from the Clearwater River and Lower Granite Reservoir typically had the largest percentage of fine-grained sediments and the highest major and trace element concentrations. Conversely, the samples from the Snake River upstream from the confluence typically had the smallest percentage of fine-grained sediments and the lowest major and trace element concentrations.

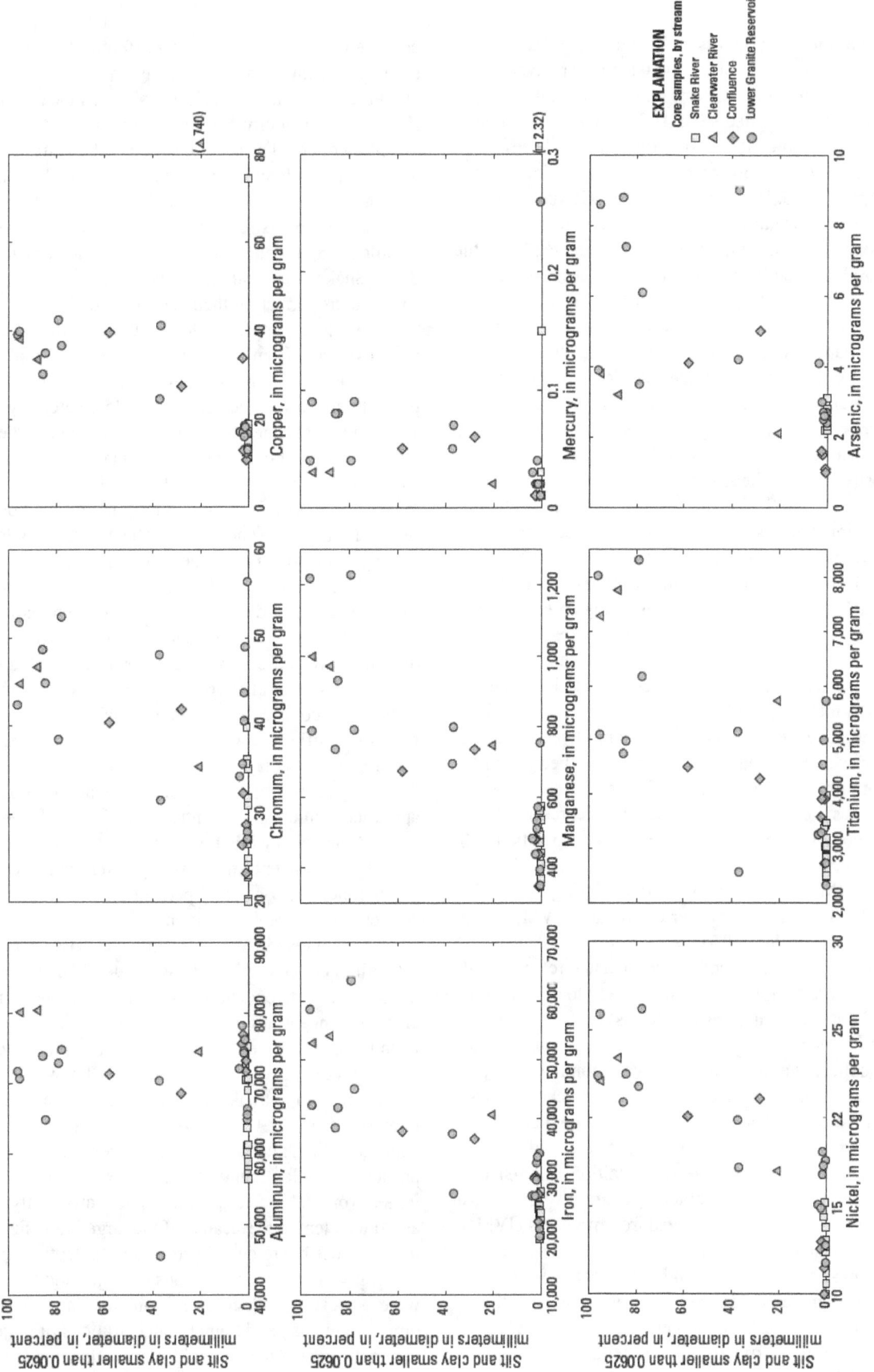

Figure 13. Comparison of percent silt and clay to selected major and trace element concentrations from bed-sediment core samples collected in Lower Granite Reservoir, the Snake and Clearwater Rivers, and the confluence of the Snake and Clearwater Rivers, eastern Washington and northern Idaho, 2010.

Summary

Lower Granite Reservoir is immediately downstream from the confluence of the Snake and Clearwater Rivers in eastern Washington and northern Idaho. According to U.S. Army Corps of Engineers (USACE), 2.6 million cubic yards (2.0 million cubic meters) of sediment have been deposited annually into Lower Granite Reservoir since its impoundment in 1975. Historically, the USACE dredged sediment to keep navigation channels clear and to maintain the flow capacity because increases in reservoir stage near the confluence reduce the effectiveness of the levees protecting the cities of Clarkston, Wash., and Lewiston, Idaho, against flooding. However, in recent years, other Federal agencies, Native American governments, and special interest groups have questioned the negative effects that dredging might have on threatened or endangered species. To help address these concerns, the U.S. Geological Survey (USGS), in cooperation with the USACE, collected bed-sediment core samples upstream from the Lower Granite Dam in the Lower Granite Reservoir and Snake and Clearwater Rivers. A total of 69 bed-sediment cores were collected at 67 locations in the study area, which includes Lower Granite Reservoir, the Snake and Clearwater Rivers, and the confluence of these two rivers, using one or more of the following corers: piston, gravity, vibrating (also called a vibracore), or box. From these 69 cores, 185 subsamples were removed and submitted for grain size analyses, 50 of which were surficial-sediment subsamples. A total of 50 subsamples were also submitted for major and trace element analyses. The multiyear project included surveys of the sedimentary structures (bedforms) of Lower Granite Reservoir using different methods. This report describes the results of a bed-sediment core samples (hereinafter cores) collected by the USGS in cooperation with the USACE during the spring and fall of 2010. A multibeam echosounding (MBES) bathymetric survey during fall 2009 and winter 2010 and an underwater video map (UVM) survey of sediment facies during fall 2009 and winter 2010, also part of the multiyear study, are briefly described and referred to in the context of core sample analyses, and used to help interpret surficial sedimentary structures (bedforms) in the study area. The grain-size distribution of surficial-sediment subsamples from cores collected underwater in Lower Granite Reservoir, the Snake and Clearwater Rivers, and the confluence of these two rivers are described, along with the down-core grain-size distribution in cores collected at or near the confluence of the Snake and Clearwater Rivers. In addition, grain-size analyses of cores are used to provide a quantitative mechanism for verifying the facies map generated from previous UVM surveys. Percent silt and clay in surficial sediment samples collected from sites at the lower end of the reservoir near the dam, where stream velocities are lower, tended to have the largest proportions (more than 80 percent) of silt and clay. Conversely, all of the surficial sediment samples collected in the Snake River upstream from the confluence had less than 20 percent silt and clay, probably a result of velocities in the Snake River high enough to keep fine-grained sediment particles entrained. Most of the surficial sediment samples collected in the Clearwater River (9 out of 13) contained less than 40 percent silt and clay. Only one site had more than 60 percent silt and clay, and this site was located in a near-shore, lower-velocity margin environment. Surficial sediment samples collected near midchannel at the confluence tended to have more silt and clay than most surficial sediment sample collection sites on the Snake and Clearwater Rivers or even sites further downstream in Lower Granite Reservoir. The turbulence and reduction in velocity induced by the confluence of the Snake and Clearwater Rivers likely caused these two rivers to drop much of their sediment load in this area.

Two core samples collected at the confluence were extracted from a thick sediment deposit that likely accumulated predominantly between 1995 and 1997; large peak streamflows occurred on the Clearwater River in both 1996 and 1997. The core collected near the middle of this sediment deposit consisted predominantly of silt and clay with some fine to very fine sand, whereas the core collected closer to the margin of this sediment deposit consisted predominantly of fine to medium sand. Cores collected on the Clearwater River just upstream from the confluence included intervals with a high percentage of silt and clay at approximately the same depths. Based on the cross section generated from the bathymetric surveys, it appears as though these silt and clay deposits may be flood-related deposits associated with large peak streamflows in 1996 and 1997 on the Clearwater River. Both of the cores collected on the Snake River just upstream from the confluence were 2 cm or less in length and neither contained silt or clay.

Fifty samples from 15 cores were analyzed for major and trace elements. Concentrations of trace elements in most of the cores were low with respect to sediment-quality guidelines. Concentrations were greater than threshold effect concentrations in 43 percent of the analyses in core 31 collected from the reservoir and 29 percent of the analyses in core 9 collected from the Clearwater River. Copper and chromium concentrations most frequently exceeded their threshold effect concentrations. Typically, major and trace element concentrations were lower in the samples collected from the Snake River upstream from the confluence as compared to those collected from the Clearwater River, the confluence of the Snake and Clearwater Rivers, and Lower Granite Reservoir. Large variations in grain size occurred with depth in the sediment cores. The percent of silt and clay-sized particles (smaller than 0.0625 millimeter [mm] in diameter) varied from 0.2 to 96.1 percent in the cores analyzed for major and trace elements. Because of the large grain size variability, it is not possible to describe trends with depth in the cores. Generally, lower concentrations of major and trace elements were associated with coarser sediments (larger than 0.0625 mm in diameter) and higher concentrations of major and trace elements were associated with finer sediments (smaller than 0.0625 mm in diameter). Samples from the Clearwater River and Lower Granite Reservoir typically had higher proportions of fine-grained sediments and higher concentrations of major and trace elements as compared to samples collected from the Snake River upstream from the confluence.

References Cited

Briggs, P.H., and Meier, A.L., 2002, The determination of forty-two elements in geological materials by inductively coupled plasma-mass spectrometry for NAWQA, *in* Taggart, J.E., Jr., ed., Analytical methods for chemical analysis of geologic and other materials: U.S. Geological Survey Open-File Report 02–223–J, p. J1–J14. (Also available at http://pubs.usgs.gov/of/2002/ofr-02-0223/J22NAWQAMethod_M.pdf.)

Emery, K.O., and Hulsemann, J., 1964, Shortening of sediment cores collected in open barrel gravity corers: Sedimentology, v. 3, p. 144–154.

Fan, Jiahua, and Morris, G.L., 1992, Reservoir sedimentation—Delta and density current deposits: Journal of Hydraulic Engineering, v. 118, no. 3, p. 354–369.

Germanoski, D., and Schumm, S.A., 1993, Changes in braided river morphology resulting from aggradation and degradation: The Journal of Geology, v. 101, no. 4, p. 451–466.

Guy, H.P., 1969, Laboratory theory and methods for sediment analysis: U.S. Geological Survey Techniques of Water-Resources Investigations, chap. Cl, book 5, 58 p.

Hageman, P.L., 2007, Determination of mercury in aqueous and geologic materials by continuous flow-cold vapor-atomic fluorescence spectrometry (CVAFS): U.S. Geological Survey Techniques and Methods, book 5, chap. D2, 6 p.

Horowitz, A.J., and Elrick, K.A., 1987, The relation of stream sediment surface area, grain size and composition to trace element chemistry: Applied Geochemistry, v. 2, p. 437–451.

Juracek, K.E., 1998, Analysis of lake-bottom sediment to estimate historical nonpoint-source phosphorous loads: Journal of the American Water Resources Association, v. 34, no. 6, p. 1449–1463.

King, J.G., Emmett, W.W., Whiting, P.J., Kenworthy, R.P., and Barry, J.J., 2004, Sediment transport data and related information for selected coarse-bed streams and rivers in Idaho: U.S. Department of Agriculture Forest Service, Rocky Mountain Research Station General Technical Report RMRS–GTR–131, 26 p., accessed July 9, 2012, at http://www.fs.fed.us/rm/boise/publications/watershed/rmrs_gtr131.pdf.

MacDonald, D.D., Ingersoll, C.G., and Berger, T.A., 2000, Development and evaluation of consensus-based sediment quality guidelines for freshwater ecosystems: Archives of Environmental Contamination and Toxicology, v. 39, p. 20–31.

Meyer, A., and Fisher, A., 1997, Data report: Grain-size analysis of sediments from the northern Barbados accretionary prism, *in* Shipley, T.H., Ogawa, Y., Blum, P., and Bahr, J.M., eds., Proceedings of the Ocean Drilling Program, Scientific Results, College Station, Tex., Ocean Drilling Program, p. 337–341.

National Agriculture Imagery Program, 2009, 1-Meter compressed county mosaic map: U.S. Department of Agriculture Farm Service Agency, Aerial Photography Field Office, scale 1:3,780.

National Institute of Standards and Technology, 2003a, Certificate of analysis, standard reference material 2709, San Joaquin soil—Baseline trace element concentrations: National Institute of Standards and Technology, accessed February 3, 2011, at https://www-s nist.gov/srmors/certificates/2709.pdf.

National Institute of Standards and Technology, 2003b, Certificate of analysis, standard reference material 2711, Montana soil—Moderately elevated trace element concentrations: National Institute of Standards and Technology, accessed June 15, 2011, at https://www-s nist.gov/srmors/certificates/view_certPDF.cfm?certificate=2711.

National Institute of Standards and Technology, 2008, Report of investigation, reference material 8704, Buffalo River sediment: National Institute of Standards and Technology, accessed February 3, 2011, at https://www-s.nist.gov/srmors/reports/8704.pdf.

Parkinson, S., Anderson, K., Conner, J., and Milligan J., 2003, Sediment transport, supply and stability in the Hells Canyon Reach of the Snake River: Idaho Power Technical Report for FERC No. 1971, 140 p.

Pirkey, K.D., and Glodt, S.R., 1998, Quality control at the U.S. Geological Survey National Water Quality Laboratory: U.S. Geological Survey Fact Sheet FS–026–98, 4 p.

Pope, L.R., and Ward, C.W., eds., 1998, Manual on test sieving methods—Guidelines for establishing sieve analysis procedures (4th ed.): West Conshohocken, Pa., American Society for Testing and Materials, 43 p.

Potts, P.J., Tindle, A.G., and Webb, P.C., 1992, Geochemical reference material compositions—Rocks, minerals, sediments, soils, carbonates, refractories, and ores used in research and industry: Boca Raton, Fla., CRC Press, 313 p.

Shelton, L.R., 1994, Guidelines for collecting and processing samples of bed sediment for analysis of trace elements and organic contaminants for the National Water Quality Assessment Program: U.S. Geological Survey Open-File Report 94–458, 20 p.

Taggart, J.E., Jr., ed., 2002, Analytical methods for chemical analysis of geologic and other materials: U.S. Geological Survey Open-File Report 02–223 [variously paged].

U.S. Army Corps of Engineers, 2002, Lower Snake River juvenile salmon migration feasibility report/environmental impact statement—Appendix T: US Army Corps of Engineers, Walla Walla District, accessed September 19, 2012, at http://www nww.usace.army.mil/portals/28/docs/environmental/lsrstudy/Appendix_T.pdf.

U.S. Army Corps of Engineers, 2003, Supplemental environmental analysis for the purposes of 2003-2004 dredging (Lower Snake and Clearwater Rivers, Washington and Idaho): U.S. Army Corps of Engineers, Walla Walla District, Washington.

U.S. Army Corps of Engineers, 2004, Framework for assessment of potential effects of dredging on sensitive fish species in San Francisco Bay: U.S. Army Corps of Engineers Final Report, August 5, 2004, 001-09170-00, San Francisco District, California.

U.S. Geological Survey, 2011, USGS water data for Idaho: National Water Information System: Web interface, accessed on September 14, 2011, at http://waterdata.usgs.gov/id/nwis.

Van Metre, P.C., Wilson, J.T., Fuller, C.C., Callender, Edward, and Mahler, B.J., 2004, Collection, analysis, and age-dating of sediment cores from 56 U.S. lakes and reservoirs sampled by the U.S. Geological Survey, 1992–2001: U.S. Geological Survey Scientific Investigations Report 2004–5184, 180 p.

Vibracoring Concepts, 2011, Core sampling basics, *in* Vibracoring a practical guide: Vibracoring Concepts, accessed on February 15, 2011, at http://www.vibracoring.com/VCconcepts html.

Wentworth, C.K., 1922, A scale of grade and class terms for clastic sediments: The Journal of Geology, v. 30, p. 377–392.

Williams, M.L., Fosness, R.L., and Weakland, R.J., 2012, Bathymetric and underwater video survey of Lower Granite Reservoir and vicinity, Washington and Idaho, 2009–10: U.S. Geological Survey Scientific Investigations Report 2012–5089, 22 p.

Appendix 1—Distance from Bank and Elevation Data for Cross Sections 17, 18, and 24

Appendix 1.   Distance from bank and elevation data for cross sections 17, 18, and 24 measured near the confluence of the Clearwater and Snake Rivers in eastern Washington and and northern Idaho, 1995 and 2008.

[dist., distance; left bank, as viewed by looking downstream; m, meters; elevation above the North American Vertical Datum of 1988; NA, not applicable; 1995 data were digitized from cross sections measured by the U.S. Army Corps of Engineers during 1995; 2008 data were from cross sections measured during 2008; data provided by the U.S. Army Corps of Engineers [Gregg Teasdale, U.S. Army Corps of Engineers, written commun., 2010])

Cross section 17				Cross section 18				Cross section 24			
1995		2008		1995		2008		1995		2008	
Dist. from left bank (m)	Elevation (m)	Dist. from left bank (m)	Elevation (m)	Dist. from left bank (m)	Elevation (m)	Dist. from left bank (m)	Elevation (m)	Dist. from left bank (m)	Elevation (m)	Dist. from left bank (m)	Elevation (m)
22.2	221.819	17.3	222.332	34.2	222.627	35.1	222.452	15.6	219.902	15.1	220.456
25.6	221.471	17.3	222.277	34.2	222.151	35.1	222.324	16.9	219.210	15.1	220.456
26.8	221.415	18.3	222.228	38.2	221.200	35.1	222.059	17.9	218.400	16.9	219.999
30.4	221.200	18.3	222.171	40.9	220.808	35.1	221.693	18.9	217.258	18.8	219.755
30.9	221.083	19.5	222.100	42.0	220.728	39.3	221.327	23.5	216.480	18.8	219.755
32.4	220.871	20.8	221.975	42.7	220.725	39.3	220.961	24.9	216.100	20.9	219.450
36.8	220.700	22.3	221.854	44.2	220.522	41.6	220.715	25.7	215.854	23.1	219.054
40.0	220.500	23.9	221.762	46.0	220.222	41.6	220.471	27.0	215.805	23.1	219.054
45.6	220.090	25.7	221.637	48.2	220.034	43.9	220.288	27.8	215.805	25.3	218.688
48.4	219.183	25.7	221.515	49.2	219.852	43.9	220.105	29.5	215.711	25.3	218.688
49.8	218.861	27.6	221.393	49.3	219.709	46.3	220.041	30.4	215.606	27.6	218.383
52.5	218.400	29.6	221.241	50.5	219.376	46.3	219.889	32.5	215.222	29.9	218.139
54.2	218.330	29.6	221.116	52.6	218.663	48.8	219.706	32.9	215.100	29.9	218.139
66.6	217.121	31.7	221.055	53.4	218.400	48.8	219.553	35.0	215.019	32.3	217.865
70.3	217.004	33.8	220.964	54.9	217.600	51.4	219.218	37.4	214.794	32.3	217.865
70.6	216.789	33.8	220.903	55.4	217.300	51.4	218.913	40.6	214.628	34.6	217.865
76.4	216.539	36.0	220.781	55.6	217.176	54.0	218.547	42.6	214.562	34.6	217.865
78.9	216.829	36.0	220.686	56.8	216.753	54.0	218.212	44.3	214.539	37.0	217.743
81.3	216.856	38.2	220.564	58.2	216.000	56.5	217.813	45.3	214.546	39.4	217.621
83.9	216.533	40.4	220.442	61.5	215.520	56.5	217.356	47.6	214.669	39.4	217.621
86.7	216.340	40.4	220.287	68.3	214.800	59.1	216.716	48.5	214.686	41.8	217.591
90.0	216.238	42.7	220.074	72.5	214.200	61.7	216.350	49.4	214.712	41.8	217.591
93.3	215.691	42.7	219.860	73.5	214.008	61.7	215.981	53.8	214.948	44.3	217.499
97.1	215.595	45.0	219.555	74.8	213.832	64.3	215.737	54.1	214.959	46.8	217.499
97.5	215.470	47.3	219.129	75.9	213.720	64.3	215.524	56.4	215.111	46.8	217.499
99.1	215.341	47.3	218.821	77.5	213.616	66.9	215.310	59.6	215.381	49.2	217.499
105.5	214.953	49.6	218.607	79.0	213.582	66.9	215.036	61.7	215.504	51.7	217.530
108.8	214.948	49.6	218.333	80.4	213.493	69.4	214.850	64.2	215.591	51.7	217.530
109.7	214.981	52.0	218.059	85.7	213.316	69.4	214.667	65.9	215.680	54.2	217.652
112.0	214.975	54.4	217.876	86.8	213.293	71.9	214.515	67.0	215.716	54.2	217.652
118.8	214.779	54.4	217.724	95.1	213.302	71.9	214.362	68.0	215.721	56.7	217.743
122.5	214.771	56.8	217.632	97.9	213.141	74.5	214.149	68.6	215.711	59.2	217.774
123.7	214.813	56.8	217.477	99.3	213.133	74.5	213.783	70.1	215.711	59.2	217.774
125.7	214.809	59.3	217.294	100.9	213.055	77.1	213.661	71.2	215.749	61.7	217.865
130.8	214.660	59.3	217.172	107.0	212.700	77.1	213.722	72.9	215.733	61.7	217.865

Appendix 1.   Distance from bank and elevation data for cross sections 17, 18, and 24 measured near the confluence of the Clearwater and Snake Rivers in eastern Washington and and northern Idaho, 1995 and 2008.—Continued

[dist., distance; left bank, as viewed by looking downstream; m, meters; elevation above the North American Vertical Datum of 1988; NA, not applicable; 1995 data were digitized from cross sections measured by the U.S. Army Corps of Engineers during 1995; 2008 data were from cross sections measured during 2008; data provided by the U.S. Army Corps of Engineers [Gregg Teasdale, U.S. Army Corps of Engineers, written commun., 2010])

Cross section 17				Cross section 18				Cross section 24			
1995		2008		1995		2008		1995		2008	
Dist. from left bank (m)	Elevation (m)	Dist. from left bank (m)	Elevation (m)	Dist. from left bank (m)	Elevation (m)	Dist. from left bank (m)	Elevation (m)	Dist. from left bank (m)	Elevation (m)	Dist. from left bank (m)	Elevation (m)
133.4	214.619	61.7	217.050	109.6	212.389	79.7	213.661	74.0	215.706	64.2	217.926
140.3	214.619	64.2	216.958	113.3	212.114	82.3	213.201	75.0	215.717	66.7	218.017
142.2	214.583	64.2	216.898	114.4	211.980	82.3	213.140	76.9	215.840	66.7	218.017
144.5	214.659	66.7	216.837	117.3	211.436	84.9	213.110	78.2	215.874	69.1	218.109
148.9	214.537	69.1	216.715	118.7	211.366	84.9	213.049	79.5	215.904	69.1	218.109
151.6	214.616	69.1	216.623	122.3	211.288	87.6	212.988	80.6	215.904	71.6	218.078
156.0	214.340	71.6	216.471	123.1	211.288	90.3	212.896	81.2	215.878	74.1	217.956
160.3	214.440	71.6	216.349	124.8	211.352	90.3	212.835	82.1	215.883	74.1	217.956
176.9	212.911	74.1	216.318	127.3	211.348	92.8	212.741	83.3	215.904	76.6	218.170
178.1	212.773	74.1	216.227	130.0	211.463	95.3	212.619	85.0	215.971	79.0	218.200
185.0	213.148	76.6	216.166	132.5	211.544	95.3	212.558	86.4	216.048	79.0	218.200
193.2	213.914	79.0	216.075	135.3	211.964	97.8	212.436	87.3	216.059	81.5	218.170
198.8	214.042	79.0	215.953	138.2	212.033	97.8	212.345	90.3	216.075	81.5	218.170
200.5	214.159	81.5	215.861	142.4	212.038	100.3	212.284	91.9	216.091	84.0	218.200
202.5	214.151	83.9	215.831	144.8	211.978	100.3	212.192	96.7	216.279	86.4	218.292
202.9	214.074	83.9	215.770	147.4	211.980	102.8	212.162	99.2	216.284	86.4	218.292
205.0	214.081	86.4	215.648	150.4	212.307	105.3	212.101	100.0	216.330	88.9	218.292
210.4	214.253	86.4	215.556	153.1	212.301	105.3	212.070	101.9	216.564	91.4	218.292
217.6	214.081	88.9	215.465	160.8	212.932	107.8	212.040	103.8	216.883	91.4	218.292
221.4	214.081	88.9	215.374	163.1	213.078	107.8	212.009	110.9	217.461	93.8	218.353
223.2	214.102	91.3	215.282	163.8	213.157	110.3	212.101	114.4	217.562	93.8	218.353
226.1	214.061	91.3	215.221	165.8	213.292	110.3	212.101	116.5	217.562	96.3	218.353
226.9	214.000	93.8	215.069	167.5	213.369	112.8	212.040	116.7	217.483	98.7	218.292
231.2	214.010	93.8	214.886	170.4	213.378	115.2	211.948	117.7	217.239	98.7	218.292
238.4	213.761	96.2	214.886	170.6	213.414	115.2	211.888	118.5	216.931	101.2	218.475
241.6	213.756	96.2	214.825	172.5	213.500	117.7	211.827	120.8	216.868	101.2	218.475
244.6	213.609	98.7	214.733	176.4	213.707	117.7	211.766	122.7	216.776	103.6	218.536
245.9	213.454	98.7	214.703	178.0	213.739	120.2	211.644	123.7	216.707	106.1	218.566
251.7	213.441	101.1	214.612	180.8	213.773	122.6	211.583	126.5	216.707	106.1	218.566
257.0	213.868	103.5	214.581	183.5	213.867	122.6	211.552	128.4	216.648	108.5	218.627
260.5	214.230	103.5	214.551	185.1	213.901	125.1	211.552	130.4	216.685	108.5	218.627
261.7	214.730	106.0	214.672	187.6	213.899	125.1	211.522	131.7	216.700	111.0	218.657
264.7	215.300	108.4	214.733	190.6	213.838	127.6	211.461	133.7	216.700	111.0	218.657
267.8	215.141	108.4	214.733	192.0	213.822	130.1	211.552	135.4	216.777	113.4	218.596
272.1	215.059	110.8	214.672	193.7	213.850	130.1	211.613	136.1	216.787	115.9	218.627

Appendix 1.   Distance from bank and elevation data for cross sections 17, 18, and 24 measured near the confluence of the Clearwater and Snake Rivers in eastern Washington and and northern Idaho, 1995 and 2008.—Continued

[dist., distance; left bank, as viewed by looking downstream; m, meters; elevation above the North American Vertical Datum of 1988; NA, not applicable; 1995 data were digitized from cross sections measured by the U.S. Army Corps of Engineers during 1995; 2008 data were from cross sections measured during 2008; data provided by the U.S. Army Corps of Engineers [Gregg Teasdale, U.S. Army Corps of Engineers, written commun., 2010])

| Cross section 17 | | | | Cross section 18 | | | | Cross section 24 | | | |
| 1995 | | 2008 | | 1995 | | 2008 | | 1995 | | 2008 | |
Dist. from left bank (m)	Elevation (m)	Dist. from left bank (m)	Elevation (m)	Dist. from left bank (m)	Elevation (m)	Dist. from left bank (m)	Elevation (m)	Dist. from left bank (m)	Elevation (m)	Dist. from left bank (m)	Elevation (m)
275.9	215.051	110.8	214.551	196.1	213.939	132.6	211.647	137.7	216.787	115.9	218.627
277.2	215.500	115.6	214.429	198.2	214.037	135.1	211.677	138.4	216.777	118.4	218.657
280.3	216.371	115.6	214.337	199.9	214.045	135.1	211.677	139.5	216.780	118.4	218.657
285.1	215.487	115.6	214.218	201.3	214.092	137.6	211.769	141.0	216.797	120.8	218.627
289.3	215.414	118.0	214.127	202.9	214.249	137.6	211.830	143.6	216.776	120.8	218.627
290.9	215.114	118.0	214.035	204.8	214.341	140.1	211.921	147.9	216.809	123.3	218.657
303.9	214.702	120.4	214.005	205.8	214.372	142.7	211.982	150.4	216.819	125.8	218.657
308.4	214.841	122.8	213.914	208.8	214.534	142.7	212.043	152.3	216.781	125.8	218.657
310.0	215.041	122.8	213.822	216.1	214.821	145.2	212.073	153.7	216.728	128.3	218.779
310.6	215.181	125.3	213.761	217.7	215.028	145.2	212.101	154.6	216.718	128.3	218.779
313.6	215.341	127.7	213.667	218.5	215.094	147.7	212.131	158.2	216.707	130.7	218.688
319.4	214.758	127.7	213.667	219.1	215.231	147.7	212.314	160.3	216.562	130.7	218.688
322.3	214.692	130.2	213.697	221.9	215.331	150.2	212.375	161.6	216.562	133.2	218.779
323.8	214.969	130.2	213.728	222.3	215.420	152.7	212.467	164.0	216.589	135.7	218.871
326.3	215.231	132.7	213.728	224.1	215.628	152.7	212.558	166.3	216.599	135.7	218.871
329.0	215.295	135.0	213.757	227.9	215.845	155.2	212.619	168.4	216.648	138.2	218.962
331.0	215.295	135.0	213.787	231.2	216.119	155.2	212.710	171.5	216.652	138.2	218.962
333.3	215.253	137.4	213.726	232.0	216.147	157.7	212.802	175.0	216.672	140.6	218.901
336.9	215.253	137.4	213.604	233.9	216.518	157.7	212.893	177.2	216.690	140.6	218.901
345.4	215.399	139.8	213.513	237.5	217.081	160.2	213.015	179.3	216.620	143.1	218.718
347.9	215.351	142.1	213.422	239.5	217.755	160.2	213.107	180.4	216.510	143.1	218.718
351.2	215.454	142.1	213.361	242.1	217.913	162.7	213.198	183.7	216.561	145.5	218.779
354.4	215.429	144.5	213.300	243.4	217.948	165.2	213.320	185.2	216.582	148.0	218.779
355.4	215.697	144.5	213.208	246.3	217.956	165.2	213.351	187.2	216.545	150.5	218.810
358.7	215.611	146.9	213.147	247.8	218.020	167.7	213.442	188.2	216.530	150.5	218.810
371.6	215.989	146.9	213.086	250.0	218.330	167.7	213.472	192.5	216.530	153.0	218.840
375.2	215.930	149.2	213.056	252.3	218.532	170.2	213.533	194.2	216.498	153.0	218.840
377.4	215.989	151.6	212.995	254.0	218.583	172.7	213.503	195.3	216.481	155.4	218.871
380.4	215.989	151.6	212.873	255.7	218.620	172.7	213.533	196.8	216.514	155.4	218.871
386.5	216.223	153.9	212.781	257.2	218.628	175.2	213.747	199.0	216.529	157.9	218.871
392.7	216.243	153.9	212.751	259.0	218.583	175.2	213.686	202.1	216.546	160.4	218.901
393.3	216.326	156.2	212.721	260.6	218.563	177.7	213.744	204.6	216.535	160.4	218.901
398.5	216.243	158.6	212.660	262.5	218.557	180.3	213.744	208.7	216.557	162.9	218.962
405.3	216.410	158.6	212.446	264.5	218.479	180.3	213.805	211.4	216.529	162.9	218.962
406.6	216.485	160.9	212.355	268.6	218.471	182.8	213.896	212.9	216.487	165.4	218.901

Appendix 1.   Distance from bank and elevation data for cross sections 17, 18, and 24 measured near the confluence of the Clearwater and Snake Rivers in eastern Washington and and northern Idaho, 1995 and 2008.—Continued

[dist., distance; left bank, as viewed by looking downstream; m, meters; elevation above the North American Vertical Datum of 1988; NA, not applicable; 1995 data were digitized from cross sections measured by the U.S. Army Corps of Engineers during 1995; 2008 data were from cross sections measured during 2008; data provided by the U.S. Army Corps of Engineers [Gregg Teasdale, U.S. Army Corps of Engineers, written commun., 2010])

| Cross section 17 | | | | Cross section 18 | | | | Cross section 24 | | | |
| 1995 | | 2008 | | 1995 | | 2008 | | 1995 | | 2008 | |
Dist. from left bank (m)	Elevation (m)	Dist. from left bank (m)	Elevation (m)	Dist. from left bank (m)	Elevation (m)	Dist. from left bank (m)	Elevation (m)	Dist. from left bank (m)	Elevation (m)	Dist. from left bank (m)	Elevation (m)
409.2	216.432	160.9	212.355	270.7	218.515	185.4	213.927	214.2	216.439	167.8	218.962
412.1	216.557	163.2	212.324	273.0	218.500	185.4	213.957	214.9	216.439	167.8	218.962
416.0	216.458	165.5	212.294	275.0	218.468	187.9	213.927	216.7	216.503	170.3	219.023
417.6	216.362	165.5	212.233	276.9	218.483	187.9	213.927	218.5	216.541	172.4	218.871
422.1	216.578	167.8	212.233	278.8	218.675	190.5	213.927	221.3	216.556	172.4	218.871
424.4	216.583	170.3	212.263	281.0	218.779	190.5	213.866	223.9	216.504	175.3	218.993
427.3	216.452	170.3	212.324	283.9	218.604	193.0	213.957	224.8	216.476	175.3	218.993
428.3	216.361	172.9	212.416	285.4	218.567	193.0	213.927	226.5	216.467	177.7	218.932
431.2	216.286	172.9	212.480	286.2	218.571	195.5	213.988	228.1	216.423	180.2	218.993
435.4	216.340	175.5	212.571	288.7	218.687	198.1	214.018	229.3	216.358	180.2	218.993
441.5	216.251	178.1	212.663	294.0	218.687	198.1	214.110	231.5	216.343	182.7	218.932
448.5	216.258	178.1	212.693	296.5	218.584	200.7	214.170	233.7	216.283	182.7	218.932
453.5	216.369	180.6	212.845	298.5	218.727	203.2	214.231	240.3	216.280	185.2	218.901
456.8	216.369	180.6	212.967	300.2	218.828	203.2	214.292	241.6	216.262	187.7	218.932
462.0	216.452	183.2	213.059	302.2	218.831	205.8	214.353	242.8	216.252	187.7	218.932
465.9	216.452	183.2	213.211	304.7	218.692	205.8	214.414	248.0	216.241	189.6	218.932
471.4	216.353	185.8	213.333	305.7	218.651	208.4	214.536	249.0	216.182	192.0	219.023
474.3	216.257	185.8	213.394	308.5	218.651	208.4	214.597	250.2	216.123	192.0	219.023
477.2	216.216	188.4	213.547	309.6	218.699	213.6	214.628	255.6	216.129	194.5	219.115
481.7	215.951	188.4	213.638	314.2	218.695	213.6	214.719	257.5	216.042	194.5	219.115
487.3	215.943	190.9	213.760	316.8	218.629	213.6	214.814	260.8	216.480	196.9	219.084
490.8	215.798	190.9	213.821	318.0	218.661	216.2	214.966	263.6	216.124	199.4	218.962
496.6	216.052	193.5	213.943	320.7	218.652	218.9	215.057	266.2	216.188	199.4	218.962
500.3	216.230	196.1	213.973	322.0	218.748	218.9	215.179	268.6	216.232	201.8	218.871
505.5	216.342	196.1	213.973	322.8	218.748	221.4	215.301	270.2	216.267	204.3	218.932
509.0	216.389	198.7	214.004	324.2	218.689	221.4	215.393	271.7	216.200	204.3	218.932
511.2	216.348	201.4	214.004	325.5	218.695	224.0	215.515	272.5	216.129	206.8	219.084
516.1	216.334	201.4	214.034	326.0	218.716	224.0	215.758	273.4	216.123	209.2	219.237
518.4	216.278	204.0	214.065	326.4	218.800	226.5	215.789	274.8	216.149	209.2	219.237
520.6	216.238	204.0	214.034	327.7	219.032	226.5	215.880	276.3	216.161	211.7	219.298
525.5	216.292	206.6	214.126	328.9	219.071	229.0	216.002	276.9	216.129	211.7	219.298
530.4	216.300	206.6	214.126	329.3	219.940	229.0	216.094	278.0	216.129	214.1	219.267
533.8	216.188	209.3	214.187	NA	NA	231.6	216.216	280.3	216.194	214.1	219.267
538.5	216.188	209.3	214.187	NA	NA	231.6	216.246	282.0	216.213	216.6	219.176
543.0	216.111	211.9	214.217	NA	NA	234.2	216.460	283.9	216.230	219.1	219.054

Appendix 1.   Distance from bank and elevation data for cross sections 17, 18, and 24 measured near the confluence of the Clearwater and Snake Rivers in eastern Washington and and northern Idaho, 1995 and 2008.—Continued

[dist., distance; left bank, as viewed by looking downstream; m, meters; elevation above the North American Vertical Datum of 1988; NA, not applicable; 1995 data were digitized from cross sections measured by the U.S. Army Corps of Engineers during 1995; 2008 data were from cross sections measured during 2008; data provided by the U.S. Army Corps of Engineers [Gregg Teasdale, U.S. Army Corps of Engineers, written commun., 2010])

Cross section 17				Cross section 18				Cross section 24			
1995		2008		1995		2008		1995		2008	
Dist. from left bank (m)	Elevation (m)	Dist. from left bank (m)	Elevation (m)	Dist. from left bank (m)	Elevation (m)	Dist. from left bank (m)	Elevation (m)	Dist. from left bank (m)	Elevation (m)	Dist. from left bank (m)	Elevation (m)
548.5	216.145	214.6	214.187	NA	NA	234.2	216.764	285.9	216.210	219.1	219.054
552.4	216.082	214.6	214.156	NA	NA	236.7	217.069	287.3	216.161	223.9	218.962
555.6	215.944	214.6	214.126	NA	NA	239.2	217.465	290.0	215.931	223.9	218.962
562.1	215.938	217.3	214.156	NA	NA	239.2	217.648	291.7	215.889	226.3	218.932
566.0	215.881	219.9	214.156	NA	NA	241.8	217.801	294.9	215.889	226.3	218.932
572.9	215.881	219.9	214.156	NA	NA	241.8	217.743	296.3	215.844	228.7	218.901
576.7	215.805	222.5	214.095	NA	NA	244.4	217.804	297.2	215.829	228.7	218.901
579.0	215.811	225.2	214.065	NA	NA	247.0	217.926	299.0	215.774	231.1	218.901
581.3	215.839	225.2	214.034	NA	NA	247.0	218.044	301.1	215.722	233.5	218.871
583.5	215.749	227.9	214.004	NA	NA	249.6	218.166	302.2	215.708	235.0	218.718
589.2	215.728	227.9	214.004	NA	NA	249.6	218.258	304.4	215.582	235.0	218.718
591.3	215.651	230.6	214.034	NA	NA	252.1	218.380	306.6	215.541	237.4	218.657
595.5	215.740	230.6	214.034	NA	NA	252.1	218.471	308.2	215.556	237.4	218.657
600.6	215.423	233.3	214.037	NA	NA	254.6	218.532	309.9	215.582	239.9	218.596
603.3	215.330	235.9	214.068	NA	NA	257.2	218.563	312.4	215.641	242.4	218.505
606.3	215.330	235.9	214.037	NA	NA	257.2	218.593	314.5	215.716	242.4	218.505
609.1	215.380	238.6	214.007	NA	NA	259.9	218.502	316.3	215.797	244.8	218.383
611.1	215.392	238.6	213.946	NA	NA	262.5	218.471	316.9	215.846	244.8	218.383
612.1	215.545	241.3	213.854	NA	NA	262.5	218.471	318.1	215.846	247.4	218.261
613.0	215.671	241.3	213.824	NA	NA	265.1	218.441	319.3	215.835	247.4	218.261
615.9	215.733	243.9	213.824	NA	NA	267.7	218.471	321.0	215.931	249.9	217.865
616.8	215.837	246.6	213.854	NA	NA	267.7	218.410	322.6	216.007	252.4	217.713
617.9	215.902	246.6	213.748	NA	NA	270.3	218.410	323.9	216.059	252.4	217.713
622.4	215.937	249.2	213.656	NA	NA	272.9	218.410	325.6	216.080	254.9	217.621
630.5	216.118	249.2	213.717	NA	NA	272.9	218.471	327.9	216.059	257.3	217.499
632.9	216.113	251.9	213.748	NA	NA	275.6	218.529	331.0	216.006	257.3	217.499
635.7	216.053	254.6	213.778	NA	NA	278.2	218.590	334.5	215.990	259.9	217.408
638.2	216.042	254.6	213.870	NA	NA	278.2	218.651	339.7	215.941	259.9	217.408
642.0	215.925	257.3	213.900	NA	NA	280.7	218.651	343.8	215.931	262.4	217.255
645.4	215.867	257.3	213.900	NA	NA	280.7	218.590	347.8	215.931	264.9	217.316
646.7	215.811	259.9	213.839	NA	NA	283.3	218.560	349.3	215.963	264.9	217.316
650.1	215.924	259.9	213.839	NA	NA	285.9	218.529	351.8	216.032	267.4	217.225
654.5	215.810	262.6	213.992	NA	NA	285.9	218.499	355.1	216.032	269.8	217.194
659.3	215.742	262.6	214.053	NA	NA	288.5	218.590	358.2	216.589	269.8	217.194
664.8	215.602	265.3	214.174	NA	NA	288.5	218.590	359.9	217.177	272.3	216.951

Appendix 1.   Distance from bank and elevation data for cross sections 17, 18, and 24 measured near the confluence of the Clearwater and Snake Rivers in eastern Washington and and northern Idaho, 1995 and 2008.—Continued

[dist., distance; left bank, as viewed by looking downstream; m, meters; elevation above the North American Vertical Datum of 1988; NA, not applicable; 1995 data were digitized from cross sections measured by the U.S. Army Corps of Engineers during 1995; 2008 data were from cross sections measured during 2008; data provided by the U.S. Army Corps of Engineers [Gregg Teasdale, U.S. Army Corps of Engineers, written commun., 2010])

| Cross section 17 | | | | Cross section 18 | | | | Cross section 24 | | | |
| 1995 | | 2008 | | 1995 | | 2008 | | 1995 | | 2008 | |
Dist. from left bank (m)	Elevation (m)	Dist. from left bank (m)	Elevation (m)	Dist. from left bank (m)	Elevation (m)	Dist. from left bank (m)	Elevation (m)	Dist. from left bank (m)	Elevation (m)	Dist. from left bank (m)	Elevation (m)
665.8	215.502	265.3	214.296	NA	NA	291.1	218.621	361.7	217.960	272.3	216.951
668.7	215.436	268.1	214.662	NA	NA	293.6	218.560	363.0	218.360	274.8	216.951
672.3	215.461	268.1	214.936	NA	NA	293.6	218.560	365.0	218.860	274.8	216.951
676.2	215.288	270.7	215.089	NA	NA	296.2	218.529	367.0	219.300	277.3	216.890
681.4	215.129	270.7	215.363	NA	NA	298.8	218.590	368.5	219.600	277.3	216.890
686.4	215.169	273.4	215.455	NA	NA	298.8	218.682	369.7	219.954	279.8	216.676
693.4	215.280	273.4	215.607	NA	NA	301.3	218.755	370.8	220.112	279.8	216.676
695.8	215.495	276.1	215.272	NA	NA	301.3	218.697	NA	NA	282.2	216.341
695.4	215.742	276.1	215.512	NA	NA	303.9	218.666	NA	NA	284.7	216.250
698.4	215.797	278.8	215.025	NA	NA	303.9	218.636	NA	NA	284.7	216.250
698.6	216.051	278.8	214.964	NA	NA	306.4	218.636	NA	NA	287.2	215.975
705.0	215.950	281.5	215.177	NA	NA	308.9	218.636	NA	NA	287.2	215.975
708.8	215.882	281.5	215.604	NA	NA	311.4	218.636	NA	NA	289.6	215.792
711.2	215.812	284.3	215.848	NA	NA	311.4	218.636	NA	NA	292.0	215.609
714.7	215.865	284.3	216.214	NA	NA	313.9	218.639	NA	NA	292.0	215.609
718.3	216.051	287.0	216.274	NA	NA	316.2	218.608	NA	NA	294.4	215.457
721.4	216.284	287.0	216.122	NA	NA	316.2	218.547	NA	NA	296.8	215.274
726.4	217.900	289.7	215.970	NA	NA	318.4	218.608	NA	NA	296.8	215.274
730.6	219.318	289.7	215.787	NA	NA	320.4	218.669	NA	NA	299.2	215.183
730.6	219.577	292.4	215.573	NA	NA	322.3	218.642	NA	NA	299.2	215.183
734.2	220.089	292.4	215.540	NA	NA	324.1	218.672	NA	NA	301.6	215.061
737.4	220.952	295.2	215.357	NA	NA	325.7	219.160	NA	NA	304.0	214.939
NA	NA	295.2	215.205	NA	NA	327.1	219.831	NA	NA	304.0	214.939
NA	NA	297.9	215.051	NA	NA	328.3	220.775	NA	NA	306.4	214.908
NA	NA	297.9	214.716	NA	NA	329.2	221.388	NA	NA	306.4	214.908
NA	NA	300.7	214.625	NA	NA	NA	NA	NA	NA	309.9	214.939
NA	NA	303.4	214.594	NA	NA	NA	NA	NA	NA	309.9	214.939
NA	NA	303.4	214.533	NA	NA	NA	NA	NA	NA	312.3	214.969
NA	NA	309.0	214.533	NA	NA	NA	NA	NA	NA	312.3	214.969
NA	NA	309.0	214.503	NA	NA	NA	NA	NA	NA	314.6	214.939
NA	NA	309.0	214.533	NA	NA	NA	NA	NA	NA	316.9	215.030
NA	NA	311.7	214.533	NA	NA	NA	NA	NA	NA	316.9	215.030
NA	NA	314.5	214.533	NA	NA	NA	NA	NA	NA	319.3	215.183
NA	NA	314.5	214.533	NA	NA	NA	NA	NA	NA	321.6	215.366
NA	NA	317.2	214.564	NA	NA	NA	NA	NA	NA	321.6	215.366

Appendix 1. Distance from bank and elevation data for cross sections 17, 18, and 24 measured near the confluence of the Clearwater and Snake Rivers in eastern Washington and and northern Idaho, 1995 and 2008.—Continued

[dist., distance; left bank, as viewed by looking downstream; m, meters; elevation above the North American Vertical Datum of 1988; NA, not applicable; 1995 data were digitized from cross sections measured by the U.S. Army Corps of Engineers during 1995; 2008 data were from cross sections measured during 2008; data provided by the U.S. Army Corps of Engineers [Gregg Teasdale, U.S. Army Corps of Engineers, written commun., 2010])

Cross section 17				Cross section 18				Cross section 24			
1995		2008		1995		2008		1995		2008	
Dist. from left bank (m)	Elevation (m)	Dist. from left bank (m)	Elevation (m)	Dist. from left bank (m)	Elevation (m)	Dist. from left bank (m)	Elevation (m)	Dist. from left bank (m)	Elevation (m)	Dist. from left bank (m)	Elevation (m)
NA	NA	320.0	214.533	NA	NA	NA	NA	NA	NA	324.0	215.701
NA	NA	320.0	214.533	NA	NA	NA	NA	NA	NA	326.3	215.670
NA	NA	322.7	214.533	NA	NA	NA	NA	NA	NA	326.3	215.670
NA	NA	322.7	214.564	NA	NA	NA	NA	NA	NA	328.7	215.548
NA	NA	325.4	214.777	NA	NA	NA	NA	NA	NA	331.0	215.609
NA	NA	325.4	214.990	NA	NA	NA	NA	NA	NA	333.3	215.640
NA	NA	328.1	215.143	NA	NA	NA	NA	NA	NA	335.7	215.731
NA	NA	328.1	215.234	NA	NA	NA	NA	NA	NA	335.7	215.731
NA	NA	330.8	215.292	NA	NA	NA	NA	NA	NA	338.1	215.823
NA	NA	333.5	215.353	NA	NA	NA	NA	NA	NA	340.4	215.853
NA	NA	333.5	215.505	NA	NA	NA	NA	NA	NA	342.7	215.853
NA	NA	336.1	215.627	NA	NA	NA	NA	NA	NA	342.7	215.853
NA	NA	336.1	215.688	NA	NA	NA	NA	NA	NA	345.0	215.823
NA	NA	336.1	215.780	NA	NA	NA	NA	NA	NA	347.3	215.884
NA	NA	338.8	215.841	NA	NA	NA	NA	NA	NA	349.6	215.914
NA	NA	341.5	215.841	NA	NA	NA	NA	NA	NA	351.9	215.975
NA	NA	341.5	215.902	NA	NA	NA	NA	NA	NA	354.2	216.006
NA	NA	344.1	215.902	NA	NA	NA	NA	NA	NA	354.2	216.006
NA	NA	346.8	215.932	NA	NA	NA	NA	NA	NA	356.5	216.341
NA	NA	346.8	215.932	NA	NA	NA	NA	NA	NA	358.7	217.012
NA	NA	349.4	215.932	NA	NA	NA	NA	NA	NA	358.7	217.012
NA	NA	352.1	215.902	NA	NA	NA	NA	NA	NA	361.0	217.987
NA	NA	352.1	215.902	NA	NA	NA	NA	NA	NA	363.1	218.475
NA	NA	354.8	215.932	NA	NA	NA	NA	NA	NA	363.1	218.475
NA	NA	354.8	215.993	NA	NA	NA	NA	NA	NA	365.1	219.480
NA	NA	357.4	215.993	NA	NA	NA	NA	NA	NA	366.9	219.755
NA	NA	357.4	216.024	NA	NA	NA	NA	NA	NA	366.9	219.755
NA	NA	360.1	215.963	NA	NA	NA	NA	NA	NA	368.5	220.456
NA	NA	360.1	215.993	NA	NA	NA	NA	NA	NA	369.8	220.700
NA	NA	362.8	216.115	NA	NA	NA	NA	NA	NA	370.7	221.279
NA	NA	362.8	216.420	NA	NA	NA	NA	NA	NA	370.7	221.279
NA	NA	365.5	216.511	NA	NA	NA	NA	NA	NA	NA	NA
NA	NA	365.5	216.511	NA	NA	NA	NA	NA	NA	NA	NA
NA	NA	368.2	216.542	NA	NA	NA	NA	NA	NA	NA	NA
NA	NA	368.2	216.755	NA	NA	NA	NA	NA	NA	NA	NA

Appendix 1.   Distance from bank and elevation data for cross sections 17, 18, and 24 measured near the confluence of the Clearwater and Snake Rivers in eastern Washington and and northern Idaho, 1995 and 2008.—Continued

[dist., distance; left bank, as viewed by looking downstream; m, meters; elevation above the North American Vertical Datum of 1988; NA, not applicable; 1995 data were digitized from cross sections measured by the U.S. Army Corps of Engineers during 1995; 2008 data were from cross sections measured during 2008; data provided by the U.S. Army Corps of Engineers [Gregg Teasdale, U.S. Army Corps of Engineers, written commun., 2010])

| Cross section 17 | | | | Cross section 18 | | | | Cross section 24 | | | |
| 1995 | | 2008 | | 1995 | | 2008 | | 1995 | | 2008 | |
Dist. from left bank (m)	Elevation (m)	Dist. from left bank (m)	Elevation (m)	Dist. from left bank (m)	Elevation (m)	Dist. from left bank (m)	Elevation (m)	Dist. from left bank (m)	Elevation (m)	Dist. from left bank (m)	Elevation (m)
NA	NA	370.8	216.908	NA	NA	NA	NA	NA	NA	NA	NA
NA	NA	370.8	216.969	NA	NA	NA	NA	NA	NA	NA	NA
NA	NA	373.6	216.999	NA	NA	NA	NA	NA	NA	NA	NA
NA	NA	373.6	217.029	NA	NA	NA	NA	NA	NA	NA	NA
NA	NA	376.3	217.090	NA	NA	NA	NA	NA	NA	NA	NA
NA	NA	376.3	217.182	NA	NA	NA	NA	NA	NA	NA	NA
NA	NA	379.0	217.273	NA	NA	NA	NA	NA	NA	NA	NA
NA	NA	379.0	217.395	NA	NA	NA	NA	NA	NA	NA	NA
NA	NA	381.7	217.517	NA	NA	NA	NA	NA	NA	NA	NA
NA	NA	381.7	217.639	NA	NA	NA	NA	NA	NA	NA	NA
NA	NA	384.4	217.731	NA	NA	NA	NA	NA	NA	NA	NA
NA	NA	384.4	217.700	NA	NA	NA	NA	NA	NA	NA	NA
NA	NA	387.2	217.761	NA	NA	NA	NA	NA	NA	NA	NA
NA	NA	387.2	217.761	NA	NA	NA	NA	NA	NA	NA	NA
NA	NA	389.9	217.700	NA	NA	NA	NA	NA	NA	NA	NA
NA	NA	389.9	217.700	NA	NA	NA	NA	NA	NA	NA	NA
NA	NA	392.5	217.883	NA	NA	NA	NA	NA	NA	NA	NA
NA	NA	395.2	218.340	NA	NA	NA	NA	NA	NA	NA	NA
NA	NA	395.2	218.401	NA	NA	NA	NA	NA	NA	NA	NA
NA	NA	397.9	218.371	NA	NA	NA	NA	NA	NA	NA	NA
NA	NA	397.9	218.310	NA	NA	NA	NA	NA	NA	NA	NA
NA	NA	400.6	218.432	NA	NA	NA	NA	NA	NA	NA	NA
NA	NA	403.3	218.767	NA	NA	NA	NA	NA	NA	NA	NA
NA	NA	403.3	218.950	NA	NA	NA	NA	NA	NA	NA	NA
NA	NA	406.0	218.950	NA	NA	NA	NA	NA	NA	NA	NA
NA	NA	406.0	218.950	NA	NA	NA	NA	NA	NA	NA	NA
NA	NA	408.6	218.950	NA	NA	NA	NA	NA	NA	NA	NA
NA	NA	408.6	218.950	NA	NA	NA	NA	NA	NA	NA	NA
NA	NA	411.3	218.950	NA	NA	NA	NA	NA	NA	NA	NA
NA	NA	414.0	219.011	NA	NA	NA	NA	NA	NA	NA	NA
NA	NA	414.0	219.011	NA	NA	NA	NA	NA	NA	NA	NA
NA	NA	416.7	219.041	NA	NA	NA	NA	NA	NA	NA	NA
NA	NA	416.7	219.041	NA	NA	NA	NA	NA	NA	NA	NA
NA	NA	419.4	219.041	NA	NA	NA	NA	NA	NA	NA	NA
NA	NA	422.1	219.011	NA	NA	NA	NA	NA	NA	NA	NA

Appendix 1.   Distance from bank and elevation data for cross sections 17, 18, and 24 measured near the confluence of the Clearwater and Snake Rivers in eastern Washington and and northern Idaho, 1995 and 2008.—Continued

[dist., distance; left bank, as viewed by looking downstream; m, meters; elevation above the North American Vertical Datum of 1988; NA, not applicable; 1995 data were digitized from cross sections measured by the U.S. Army Corps of Engineers during 1995; 2008 data were from cross sections measured during 2008; data provided by the U.S. Army Corps of Engineers [Gregg Teasdale, U.S. Army Corps of Engineers, written commun., 2010])

Cross section 17				Cross section 18				Cross section 24			
1995		2008		1995		2008		1995		2008	
Dist. from left bank (m)	Elevation (m)	Dist. from left bank (m)	Elevation (m)	Dist. from left bank (m)	Elevation (m)	Dist. from left bank (m)	Elevation (m)	Dist. from left bank (m)	Elevation (m)	Dist. from left bank (m)	Elevation (m)
NA	NA	422.1	218.922	NA	NA	NA	NA	NA	NA	NA	NA
NA	NA	424.7	218.861	NA	NA	NA	NA	NA	NA	NA	NA
NA	NA	424.7	218.922	NA	NA	NA	NA	NA	NA	NA	NA
NA	NA	427.4	219.044	NA	NA	NA	NA	NA	NA	NA	NA
NA	NA	427.4	219.288	NA	NA	NA	NA	NA	NA	NA	NA
NA	NA	430.1	219.349	NA	NA	NA	NA	NA	NA	NA	NA
NA	NA	432.8	219.349	NA	NA	NA	NA	NA	NA	NA	NA
NA	NA	432.8	219.379	NA	NA	NA	NA	NA	NA	NA	NA
NA	NA	435.5	219.349	NA	NA	NA	NA	NA	NA	NA	NA
NA	NA	435.5	219.319	NA	NA	NA	NA	NA	NA	NA	NA
NA	NA	438.2	219.319	NA	NA	NA	NA	NA	NA	NA	NA
NA	NA	440.9	219.319	NA	NA	NA	NA	NA	NA	NA	NA
NA	NA	440.9	219.352	NA	NA	NA	NA	NA	NA	NA	NA
NA	NA	443.6	219.291	NA	NA	NA	NA	NA	NA	NA	NA
NA	NA	446.3	219.047	NA	NA	NA	NA	NA	NA	NA	NA
NA	NA	446.3	218.986	NA	NA	NA	NA	NA	NA	NA	NA
NA	NA	446.3	219.047	NA	NA	NA	NA	NA	NA	NA	NA
NA	NA	449.0	219.108	NA	NA	NA	NA	NA	NA	NA	NA
NA	NA	451.7	219.352	NA	NA	NA	NA	NA	NA	NA	NA
NA	NA	451.7	219.413	NA	NA	NA	NA	NA	NA	NA	NA
NA	NA	454.4	219.383	NA	NA	NA	NA	NA	NA	NA	NA
NA	NA	454.4	219.352	NA	NA	NA	NA	NA	NA	NA	NA
NA	NA	457.1	219.230	NA	NA	NA	NA	NA	NA	NA	NA
NA	NA	457.1	219.108	NA	NA	NA	NA	NA	NA	NA	NA
NA	NA	459.8	219.108	NA	NA	NA	NA	NA	NA	NA	NA
NA	NA	459.8	219.108	NA	NA	NA	NA	NA	NA	NA	NA
NA	NA	462.5	219.107	NA	NA	NA	NA	NA	NA	NA	NA
NA	NA	465.2	219.107	NA	NA	NA	NA	NA	NA	NA	NA
NA	NA	465.2	219.168	NA	NA	NA	NA	NA	NA	NA	NA
NA	NA	467.9	219.168	NA	NA	NA	NA	NA	NA	NA	NA
NA	NA	470.6	219.229	NA	NA	NA	NA	NA	NA	NA	NA
NA	NA	470.6	219.290	NA	NA	NA	NA	NA	NA	NA	NA
NA	NA	473.3	219.412	NA	NA	NA	NA	NA	NA	NA	NA
NA	NA	476.0	219.476	NA	NA	NA	NA	NA	NA	NA	NA
NA	NA	476.0	219.476	NA	NA	NA	NA	NA	NA	NA	NA

Appendix 1.   Distance from bank and elevation data for cross sections 17, 18, and 24 measured near the confluence of the Clearwater and Snake Rivers in eastern Washington and and northern Idaho, 1995 and 2008.—Continued

[dist., distance; left bank, as viewed by looking downstream; m, meters; elevation above the North American Vertical Datum of 1988; NA, not applicable; 1995 data were digitized from cross sections measured by the U.S. Army Corps of Engineers during 1995; 2008 data were from cross sections measured during 2008; data provided by the U.S. Army Corps of Engineers [Gregg Teasdale, U.S. Army Corps of Engineers, written commun., 2010])

| Cross section 17 | | | | Cross section 18 | | | | Cross section 24 | | | |
| 1995 | | 2008 | | 1995 | | 2008 | | 1995 | | 2008 | |
Dist. from left bank (m)	Elevation (m)	Dist. from left bank (m)	Elevation (m)	Dist. from left bank (m)	Elevation (m)	Dist. from left bank (m)	Elevation (m)	Dist. from left bank (m)	Elevation (m)	Dist. from left bank (m)	Elevation (m)
NA	NA	478.6	219.446	NA	NA	NA	NA	NA	NA	NA	NA
NA	NA	478.6	219.415	NA	NA	NA	NA	NA	NA	NA	NA
NA	NA	481.3	219.415	NA	NA	NA	NA	NA	NA	NA	NA
NA	NA	481.3	219.415	NA	NA	NA	NA	NA	NA	NA	NA
NA	NA	484.0	219.415	NA	NA	NA	NA	NA	NA	NA	NA
NA	NA	486.7	219.415	NA	NA	NA	NA	NA	NA	NA	NA
NA	NA	486.7	219.446	NA	NA	NA	NA	NA	NA	NA	NA
NA	NA	489.4	219.446	NA	NA	NA	NA	NA	NA	NA	NA
NA	NA	489.4	219.415	NA	NA	NA	NA	NA	NA	NA	NA
NA	NA	492.0	219.415	NA	NA	NA	NA	NA	NA	NA	NA
NA	NA	492.0	219.354	NA	NA	NA	NA	NA	NA	NA	NA
NA	NA	494.7	219.293	NA	NA	NA	NA	NA	NA	NA	NA
NA	NA	497.4	219.232	NA	NA	NA	NA	NA	NA	NA	NA
NA	NA	497.4	219.141	NA	NA	NA	NA	NA	NA	NA	NA
NA	NA	500.1	219.110	NA	NA	NA	NA	NA	NA	NA	NA
NA	NA	502.8	219.141	NA	NA	NA	NA	NA	NA	NA	NA
NA	NA	502.8	219.232	NA	NA	NA	NA	NA	NA	NA	NA
NA	NA	505.4	219.385	NA	NA	NA	NA	NA	NA	NA	NA
NA	NA	505.4	219.385	NA	NA	NA	NA	NA	NA	NA	NA
NA	NA	508.1	219.385	NA	NA	NA	NA	NA	NA	NA	NA
NA	NA	510.8	219.324	NA	NA	NA	NA	NA	NA	NA	NA
NA	NA	510.8	219.263	NA	NA	NA	NA	NA	NA	NA	NA
NA	NA	513.5	219.171	NA	NA	NA	NA	NA	NA	NA	NA
NA	NA	513.5	219.110	NA	NA	NA	NA	NA	NA	NA	NA
NA	NA	516.2	219.049	NA	NA	NA	NA	NA	NA	NA	NA
NA	NA	516.2	219.202	NA	NA	NA	NA	NA	NA	NA	NA
NA	NA	518.9	219.293	NA	NA	NA	NA	NA	NA	NA	NA
NA	NA	518.9	219.232	NA	NA	NA	NA	NA	NA	NA	NA
NA	NA	521.6	219.171	NA	NA	NA	NA	NA	NA	NA	NA
NA	NA	524.3	219.141	NA	NA	NA	NA	NA	NA	NA	NA
NA	NA	524.3	219.049	NA	NA	NA	NA	NA	NA	NA	NA
NA	NA	527.0	219.019	NA	NA	NA	NA	NA	NA	NA	NA
NA	NA	529.7	219.229	NA	NA	NA	NA	NA	NA	NA	NA
NA	NA	529.7	219.412	NA	NA	NA	NA	NA	NA	NA	NA
NA	NA	529.7	219.412	NA	NA	NA	NA	NA	NA	NA	NA

Appendix 1.   Distance from bank and elevation data for cross sections 17, 18, and 24 measured near the confluence of the Clearwater and Snake Rivers in eastern Washington and and northern Idaho, 1995 and 2008.—Continued

[dist., distance; left bank, as viewed by looking downstream; m, meters; elevation above the North American Vertical Datum of 1988; NA, not applicable; 1995 data were digitized from cross sections measured by the U.S. Army Corps of Engineers during 1995; 2008 data were from cross sections measured during 2008; data provided by the U.S. Army Corps of Engineers [Gregg Teasdale, U.S. Army Corps of Engineers, written commun., 2010])

| Cross section 17 | | | | Cross section 18 | | | | Cross section 24 | | | |
| 1995 | | 2008 | | 1995 | | 2008 | | 1995 | | 2008 | |
Dist. from left bank (m)	Elevation (m)	Dist. from left bank (m)	Elevation (m)	Dist. from left bank (m)	Elevation (m)	Dist. from left bank (m)	Elevation (m)	Dist. from left bank (m)	Elevation (m)	Dist. from left bank (m)	Elevation (m)
NA	NA	532.4	219.351	NA	NA	NA	NA	NA	NA	NA	NA
NA	NA	535.1	219.260	NA	NA	NA	NA	NA	NA	NA	NA
NA	NA	535.1	219.229	NA	NA	NA	NA	NA	NA	NA	NA
NA	NA	537.8	219.168	NA	NA	NA	NA	NA	NA	NA	NA
NA	NA	537.8	219.077	NA	NA	NA	NA	NA	NA	NA	NA
NA	NA	540.5	219.016	NA	NA	NA	NA	NA	NA	NA	NA
NA	NA	540.5	218.894	NA	NA	NA	NA	NA	NA	NA	NA
NA	NA	543.2	219.138	NA	NA	NA	NA	NA	NA	NA	NA
NA	NA	545.8	219.199	NA	NA	NA	NA	NA	NA	NA	NA
NA	NA	545.8	219.168	NA	NA	NA	NA	NA	NA	NA	NA
NA	NA	548.5	219.229	NA	NA	NA	NA	NA	NA	NA	NA
NA	NA	548.5	219.199	NA	NA	NA	NA	NA	NA	NA	NA
NA	NA	551.2	219.168	NA	NA	NA	NA	NA	NA	NA	NA
NA	NA	551.2	219.077	NA	NA	NA	NA	NA	NA	NA	NA
NA	NA	553.9	218.955	NA	NA	NA	NA	NA	NA	NA	NA
NA	NA	556.5	218.803	NA	NA	NA	NA	NA	NA	NA	NA
NA	NA	556.5	218.711	NA	NA	NA	NA	NA	NA	NA	NA
NA	NA	559.1	218.650	NA	NA	NA	NA	NA	NA	NA	NA
NA	NA	561.8	218.955	NA	NA	NA	NA	NA	NA	NA	NA
NA	NA	561.8	219.077	NA	NA	NA	NA	NA	NA	NA	NA
NA	NA	564.4	219.046	NA	NA	NA	NA	NA	NA	NA	NA
NA	NA	564.4	218.985	NA	NA	NA	NA	NA	NA	NA	NA
NA	NA	567.1	218.894	NA	NA	NA	NA	NA	NA	NA	NA
NA	NA	567.1	218.891	NA	NA	NA	NA	NA	NA	NA	NA
NA	NA	569.7	218.830	NA	NA	NA	NA	NA	NA	NA	NA
NA	NA	572.3	218.739	NA	NA	NA	NA	NA	NA	NA	NA
NA	NA	572.3	218.708	NA	NA	NA	NA	NA	NA	NA	NA
NA	NA	574.9	218.708	NA	NA	NA	NA	NA	NA	NA	NA
NA	NA	577.5	218.647	NA	NA	NA	NA	NA	NA	NA	NA
NA	NA	577.5	218.525	NA	NA	NA	NA	NA	NA	NA	NA
NA	NA	580.1	218.373	NA	NA	NA	NA	NA	NA	NA	NA
NA	NA	580.1	218.251	NA	NA	NA	NA	NA	NA	NA	NA
NA	NA	582.7	218.159	NA	NA	NA	NA	NA	NA	NA	NA
NA	NA	585.2	218.129	NA	NA	NA	NA	NA	NA	NA	NA
NA	NA	585.2	218.068	NA	NA	NA	NA	NA	NA	NA	NA

Appendix 1.   Distance from bank and elevation data for cross sections 17, 18, and 24 measured near the confluence of the Clearwater and Snake Rivers in eastern Washington and and northern Idaho, 1995 and 2008.—Continued

[dist., distance; left bank, as viewed by looking downstream; m, meters; elevation above the North American Vertical Datum of 1988; NA, not applicable; 1995 data were digitized from cross sections measured by the U.S. Army Corps of Engineers during 1995; 2008 data were from cross sections measured during 2008; data provided by the U.S. Army Corps of Engineers [Gregg Teasdale, U.S. Army Corps of Engineers, written commun., 2010])

| Cross section 17 | | | | Cross section 18 | | | | Cross section 24 | | | |
| 1995 | | 2008 | | 1995 | | 2008 | | 1995 | | 2008 | |
Dist. from left bank (m)	Elevation (m)	Dist. from left bank (m)	Elevation (m)	Dist. from left bank (m)	Elevation (m)	Dist. from left bank (m)	Elevation (m)	Dist. from left bank (m)	Elevation (m)	Dist. from left bank (m)	Elevation (m)
NA	NA	587.8	218.037	NA	NA	NA	NA	NA	NA	NA	NA
NA	NA	590.3	217.977	NA	NA	NA	NA	NA	NA	NA	NA
NA	NA	590.3	217.885	NA	NA	NA	NA	NA	NA	NA	NA
NA	NA	592.8	217.824	NA	NA	NA	NA	NA	NA	NA	NA
NA	NA	592.8	217.733	NA	NA	NA	NA	NA	NA	NA	NA
NA	NA	595.3	217.702	NA	NA	NA	NA	NA	NA	NA	NA
NA	NA	595.3	217.611	NA	NA	NA	NA	NA	NA	NA	NA
NA	NA	597.8	217.550	NA	NA	NA	NA	NA	NA	NA	NA
NA	NA	600.3	217.611	NA	NA	NA	NA	NA	NA	NA	NA
NA	NA	600.3	217.580	NA	NA	NA	NA	NA	NA	NA	NA
NA	NA	602.8	217.489	NA	NA	NA	NA	NA	NA	NA	NA
NA	NA	602.8	217.367	NA	NA	NA	NA	NA	NA	NA	NA
NA	NA	605.3	217.275	NA	NA	NA	NA	NA	NA	NA	NA
NA	NA	605.3	217.275	NA	NA	NA	NA	NA	NA	NA	NA
NA	NA	607.8	217.489	NA	NA	NA	NA	NA	NA	NA	NA
NA	NA	607.8	217.489	NA	NA	NA	NA	NA	NA	NA	NA
NA	NA	610.3	217.397	NA	NA	NA	NA	NA	NA	NA	NA
NA	NA	612.8	217.306	NA	NA	NA	NA	NA	NA	NA	NA
NA	NA	612.8	217.154	NA	NA	NA	NA	NA	NA	NA	NA
NA	NA	615.3	217.001	NA	NA	NA	NA	NA	NA	NA	NA
NA	NA	615.3	216.757	NA	NA	NA	NA	NA	NA	NA	NA
NA	NA	617.8	216.574	NA	NA	NA	NA	NA	NA	NA	NA
NA	NA	620.3	216.544	NA	NA	NA	NA	NA	NA	NA	NA
NA	NA	620.3	216.513	NA	NA	NA	NA	NA	NA	NA	NA
NA	NA	622.7	216.178	NA	NA	NA	NA	NA	NA	NA	NA
NA	NA	625.2	216.453	NA	NA	NA	NA	NA	NA	NA	NA
NA	NA	625.2	216.574	NA	NA	NA	NA	NA	NA	NA	NA
NA	NA	630.1	216.574	NA	NA	NA	NA	NA	NA	NA	NA
NA	NA	630.1	216.482	NA	NA	NA	NA	NA	NA	NA	NA
NA	NA	632.6	216.330	NA	NA	NA	NA	NA	NA	NA	NA
NA	NA	635.0	216.208	NA	NA	NA	NA	NA	NA	NA	NA
NA	NA	635.0	216.086	NA	NA	NA	NA	NA	NA	NA	NA
NA	NA	637.4	215.933	NA	NA	NA	NA	NA	NA	NA	NA
NA	NA	637.4	215.842	NA	NA	NA	NA	NA	NA	NA	NA
NA	NA	639.9	215.781	NA	NA	NA	NA	NA	NA	NA	NA

Appendix 1. Distance from bank and elevation data for cross sections 17, 18, and 24 measured near the confluence of the Clearwater and Snake Rivers in eastern Washington and and northern Idaho, 1995 and 2008.—Continued

[dist., distance; left bank, as viewed by looking downstream; m, meters; elevation above the North American Vertical Datum of 1988; NA, not applicable; 1995 data were digitized from cross sections measured by the U.S. Army Corps of Engineers during 1995; 2008 data were from cross sections measured during 2008; data provided by the U.S. Army Corps of Engineers [Gregg Teasdale, U.S. Army Corps of Engineers, written commun., 2010])

Cross section 17				Cross section 18				Cross section 24			
1995		2008		1995		2008		1995		2008	
Dist. from left bank (m)	Elevation (m)	Dist. from left bank (m)	Elevation (m)	Dist. from left bank (m)	Elevation (m)	Dist. from left bank (m)	Elevation (m)	Dist. from left bank (m)	Elevation (m)	Dist. from left bank (m)	Elevation (m)
NA	NA	642.3	215.933	NA	NA	NA	NA	NA	NA	NA	NA
NA	NA	642.3	215.994	NA	NA	NA	NA	NA	NA	NA	NA
NA	NA	644.7	215.964	NA	NA	NA	NA	NA	NA	NA	NA
NA	NA	647.1	215.994	NA	NA	NA	NA	NA	NA	NA	NA
NA	NA	647.1	216.025	NA	NA	NA	NA	NA	NA	NA	NA
NA	NA	649.6	215.994	NA	NA	NA	NA	NA	NA	NA	NA
NA	NA	649.6	215.964	NA	NA	NA	NA	NA	NA	NA	NA
NA	NA	652.0	215.872	NA	NA	NA	NA	NA	NA	NA	NA
NA	NA	652.0	215.751	NA	NA	NA	NA	NA	NA	NA	NA
NA	NA	654.4	215.598	NA	NA	NA	NA	NA	NA	NA	NA
NA	NA	656.9	215.385	NA	NA	NA	NA	NA	NA	NA	NA
NA	NA	656.9	215.415	NA	NA	NA	NA	NA	NA	NA	NA
NA	NA	656.9	215.415	NA	NA	NA	NA	NA	NA	NA	NA
NA	NA	661.9	215.537	NA	NA	NA	NA	NA	NA	NA	NA
NA	NA	661.9	215.659	NA	NA	NA	NA	NA	NA	NA	NA
NA	NA	661.9	215.629	NA	NA	NA	NA	NA	NA	NA	NA
NA	NA	664.3	215.537	NA	NA	NA	NA	NA	NA	NA	NA
NA	NA	666.8	215.443	NA	NA	NA	NA	NA	NA	NA	NA
NA	NA	666.8	215.321	NA	NA	NA	NA	NA	NA	NA	NA
NA	NA	669.3	215.229	NA	NA	NA	NA	NA	NA	NA	NA
NA	NA	669.3	215.138	NA	NA	NA	NA	NA	NA	NA	NA
NA	NA	671.7	214.985	NA	NA	NA	NA	NA	NA	NA	NA
NA	NA	674.2	214.894	NA	NA	NA	NA	NA	NA	NA	NA
NA	NA	674.2	214.772	NA	NA	NA	NA	NA	NA	NA	NA
NA	NA	676.6	214.711	NA	NA	NA	NA	NA	NA	NA	NA
NA	NA	676.6	214.589	NA	NA	NA	NA	NA	NA	NA	NA
NA	NA	679.1	214.498	NA	NA	NA	NA	NA	NA	NA	NA
NA	NA	681.6	214.437	NA	NA	NA	NA	NA	NA	NA	NA
NA	NA	681.6	214.315	NA	NA	NA	NA	NA	NA	NA	NA
NA	NA	684.1	214.196	NA	NA	NA	NA	NA	NA	NA	NA
NA	NA	686.6	214.105	NA	NA	NA	NA	NA	NA	NA	NA
NA	NA	686.6	214.074	NA	NA	NA	NA	NA	NA	NA	NA
NA	NA	689.1	214.196	NA	NA	NA	NA	NA	NA	NA	NA
NA	NA	691.6	214.288	NA	NA	NA	NA	NA	NA	NA	NA
NA	NA	691.6	214.288	NA	NA	NA	NA	NA	NA	NA	NA

Appendix 1.   Distance from bank and elevation data for cross sections 17, 18, and 24 measured near the confluence of the Clearwater and Snake Rivers in eastern Washington and and northern Idaho, 1995 and 2008.—Continued

[dist., distance; left bank, as viewed by looking downstream; m, meters; elevation above the North American Vertical Datum of 1988; NA, not applicable; 1995 data were digitized from cross sections measured by the U.S. Army Corps of Engineers during 1995; 2008 data were from cross sections measured during 2008; data provided by the U.S. Army Corps of Engineers [Gregg Teasdale, U.S. Army Corps of Engineers, written commun., 2010])

Cross section 17				Cross section 18				Cross section 24			
1995		2008		1995		2008		1995		2008	
Dist. from left bank (m)	Elevation (m)	Dist. from left bank (m)	Elevation (m)	Dist. from left bank (m)	Elevation (m)	Dist. from left bank (m)	Elevation (m)	Dist. from left bank (m)	Elevation (m)	Dist. from left bank (m)	Elevation (m)
NA	NA	694.1	214.288	NA	NA	NA	NA	NA	NA	NA	NA
NA	NA	696.5	214.653	NA	NA	NA	NA	NA	NA	NA	NA
NA	NA	696.5	214.958	NA	NA	NA	NA	NA	NA	NA	NA
NA	NA	699.0	215.354	NA	NA	NA	NA	NA	NA	NA	NA
NA	NA	699.0	215.507	NA	NA	NA	NA	NA	NA	NA	NA
NA	NA	701.5	215.751	NA	NA	NA	NA	NA	NA	NA	NA
NA	NA	701.5	215.933	NA	NA	NA	NA	NA	NA	NA	NA
NA	NA	704.0	215.903	NA	NA	NA	NA	NA	NA	NA	NA
NA	NA	706.5	215.872	NA	NA	NA	NA	NA	NA	NA	NA
NA	NA	706.5	215.812	NA	NA	NA	NA	NA	NA	NA	NA
NA	NA	708.9	215.690	NA	NA	NA	NA	NA	NA	NA	NA
NA	NA	711.4	215.568	NA	NA	NA	NA	NA	NA	NA	NA
NA	NA	713.9	215.568	NA	NA	NA	NA	NA	NA	NA	NA
NA	NA	713.9	215.632	NA	NA	NA	NA	NA	NA	NA	NA
NA	NA	713.9	215.601	NA	NA	NA	NA	NA	NA	NA	NA
NA	NA	718.8	215.601	NA	NA	NA	NA	NA	NA	NA	NA
NA	NA	718.8	215.632	NA	NA	NA	NA	NA	NA	NA	NA
NA	NA	721.2	215.662	NA	NA	NA	NA	NA	NA	NA	NA
NA	NA	723.7	215.754	NA	NA	NA	NA	NA	NA	NA	NA
NA	NA	726.0	216.028	NA	NA	NA	NA	NA	NA	NA	NA
NA	NA	726.0	216.180	NA	NA	NA	NA	NA	NA	NA	NA
NA	NA	728.4	216.485	NA	NA	NA	NA	NA	NA	NA	NA
NA	NA	728.4	216.820	NA	NA	NA	NA	NA	NA	NA	NA
NA	NA	730.6	217.369	NA	NA	NA	NA	NA	NA	NA	NA
NA	NA	732.7	217.704	NA	NA	NA	NA	NA	NA	NA	NA
NA	NA	732.7	218.314	NA	NA	NA	NA	NA	NA	NA	NA
NA	NA	734.7	218.893	NA	NA	NA	NA	NA	NA	NA	NA
NA	NA	736.5	219.106	NA	NA	NA	NA	NA	NA	NA	NA
NA	NA	738.2	219.536	NA	NA	NA	NA	NA	NA	NA	NA
NA	NA	738.2	219.871	NA	NA	NA	NA	NA	NA	NA	NA
NA	NA	739.7	220.329	NA	NA	NA	NA	NA	NA	NA	NA
NA	NA	741.0	220.938	NA	NA	NA	NA	NA	NA	NA	NA

Appendix 2—Grain-Size Data from Bed-Sediment Core Samples

Appendix 2.   Grain-size data from bed-sediment core samples collected in Lower Granite Reservoir and the Clearwater and Snake Rivers just above their confluence, 2010.

[# , number; cm, centimeter; mm, millimeter; --, not analyzed]

USGS station number	Core identifier	Core sample interval analyzed (cm)	Sample identifier	Date	Percent finer than 16 mm; coarse/ medium pebbles	Percent finer than 8 mm; me-dium/fine pebbles	Percent finer than 4 mm; fine/ very fine pebbles	Percent finer than 2 mm; very fine pebbles/ very coarse sand	Percent finer than 1 mm; very coarse sand/ coarse sand	Percent finer than 0.50 mm; coarse/ medium sand
				Clearwater River						
462534117015500	1	0–5	Core #1 0–5	5/14/10	--	--	--	100.0	97.9	94.9
462534117015500	1	9–12	Core #1 9–12	5/14/10	--	--	--	--	99.7	94.2
462537117015600	2	60–63	Core #2 60–63	5/13/10	--	--	--	--	100.0	98.0
462537117015600	2	80–83	Core #2 80–83	5/13/10	--	--	--	100.0	99.9	99.9
462537117015600	2	105–108	Core #2 105–108	5/13/10	--	--	--	--	100.0	99.7
462535117015700	3	2–5	Core #3 2–5	5/14/10	--	--	--	100.0	99.9	98.2
462535117015700	3	28–31	Core #3 28–31	5/14/10	--	--	--	100.0	99.8	98.5
462535117015700	3	62–65	Core #3 62–65	5/14/10	--	--	--	--	--	100.0
462535117015700	3	90–93	Core #3 90–93	5/14/10	--	--	--	--	100.0	99.9
462535117015700	3	114–117	Core #3 114–117	5/14/10	--	--	--	--	--	100.0
462535117015700	3	123–126	Core #3 123–126	5/14/10	--	--	--	--	100.0	99.5
462532117014900	4	22–25	Core #4 22–25	5/13/10	--	--	--	100.0	99.7	94.8
462532117014900	4	44–47	Core #4 44–47	5/13/10	--	--	--	100.0	99.8	94.4
462535117014800	5	0–2	Core #5 0–2	5/14/10	--	--	--	100.0	99.6	97.8
462535117014800	5	4–6	Core #5 4–6	5/14/10	--	--	--	100.0	99.7	96.6
462535117014800	5	20–22	Core #5 20–22	5/14/10	--	--	--	100.0	99.8	96.0
462535117014800	5	43–45	Core #5 43–45	5/14/10	--	--	--	100.0	99.2	96.7
462527117011300	7	0–2	Core #7 0–2	5/18/10	--	100.0	99.5	98.7	98.3	76.4
462532117011101	9	18–21	Core #9 18–21	5/13/10	--	--	--	--	100.0	99.9
462532117011101	9	34–37	Core #9 34–37	5/13/10	--	--	--	--	100.0	99.7
462532117011101	9	50–53	Core #9 50–53	5/13/10	--	--	--	--	--	100.0
462517117010300	10	0–1	Core #10 0–1	5/13/10	--	--	--	100.0	99.7	90.2
462517117010300	10	20–23	Core #10 20–23	5/13/10	--	--	100.0	99.6	98.5	67.8
462517117010300	10	80–83	Core #10 80–83	5/13/10	--	--	--	--	100.0	99.3
462517117010300	10	108–110	Core #10 108–110	5/13/10	--	100.0	99.6	99.3	97.9	47.8
462520117005900	11	0–3	Core #11 0–3	5/18/10	--	--	--	100.0	99.9	81.8
462524117005900	12	0–2	Core #12 0–2	5/13/10	--	--	--	100.0	99.0	91.7
462524117005900	12	29–32	Core #12 29–32	5/13/10	--	--	--	100.0	99.9	91.3
462513117004900	13	0–4	Core #13 0–4	5/18/10	--	--	100.0	99.6	98.5	50.5
462517117004900	14	0–1	Core #14 0–1	5/18/10	--	--	--	100.0	99.6	84.5
462517117004900	14	2–5	Core #14 2–5	5/18/10	--	--	--	100.0	99.8	68.9
462517117004900	14	54–56	Core #14 54–56	5/18/10	--	--	--	100.0	99.9	98.0
462520117004700	15	2–5	Core #15 2–5	5/18/10	--	--	--	--	100.0	87.8
462520117004700	15	17–20	Core #15 17–20	5/18/10	--	--	--	100.0	99.3	80.5
462520117004700	15	22–25	Core #15 22–25	5/18/10	--	--	--	100.0	99.6	91.5

Appendix 2.   Grain-size data from bed-sediment core samples collected in Lower Granite Reservoir and the Clearwater and Snake Rivers just above their confluence, 2010.—Continued

[# , number; cm, centimeter; mm, millimeter; --, not analyzed]

Sample identifier	Date	Percent finer than 0.25 mm; medium/ fine sand	Percent finer than 0.125 mm; fine/very fine sand	Percent finer than 0.0625 mm; very fine sand/ coarse silt	Percent finer than 0.031 mm; coarse/ medium silt	Percent finer than 0.016 mm; medium/ fine silt	Percent finer than 0.008 mm; fine/very fine silt	Percent finer than 0.004 mm; very fine silt/clay	Percent finer than 0.002 mm; silt/clay for mineral analysis	Percent finer than 0.001 mm; finer than clay
					Clearwater River					
Core #1 0–5	5/14/10	62.2	48.2	17.2	16.1	12.8	9.8	7.6	5.8	3.6
Core #1 9–12	5/14/10	16.9	1.4	0.7	--	--	--	--	--	--
Core #2 60–63	5/13/10	40.5	3.4	1.2	--	--	--	--	--	--
Core #2 80–83	5/13/10	99.2	97.8	83.0	55.1	32.3	24.8	19.3	15.5	11.3
Core #2 105–108	5/13/10	93.8	12.6	3.1	--	--	--	--	--	--
Core #3 2–5	5/14/10	48.7	2.3	0.6	--	--	--	--	--	--
Core #3 28–31	5/14/10	62.2	10.7	3.0	--	--	--	--	--	--
Core #3 62–65	5/14/10	99.9	99.5	94.0	70.7	45.7	34.4	27.0	21.5	17.8
Core #3 90–93	5/14/10	95.1	14.7	2.8	--	--	--	--	--	--
Core #3 114–117	5/14/10	99.8	97.5	90.2	72.0	45.8	34.4	25.8	21.1	20.3
Core #3 123–126	5/14/10	97.1	59.5	38.1	29.4	19.6	14.6	11.2	8.4	7.9
Core #4 22–25	5/13/10	8.9	1.2	0.6	--	--	--	--	--	--
Core #4 44–47	5/13/10	20.0	2.9	1.8	--	--	--	--	--	--
Core #5 0–2	5/14/10	75.4	61.4	53.1	49.6	40.6	30.9	23.8	18.7	11.7
Core #5 4–6	5/14/10	27.0	1.5	0.6	--	--	--	--	--	--
Core #5 20–22	5/14/10	25.1	1.6	0.6	--	--	--	--	--	--
Core #5 43–45	5/14/10	48.9	8.3	2.5	--	--	--	--	--	--
Core #7 0–2	5/18/10	23.2	8.5	5.4	5.0	3.8	2.6	1.9	1.4	0.6
Core #9 18–21	5/13/10	98.8	95.7	87.7	66.4	42.5	32.4	25.5	20.5	14.2
Core #9 34–37	5/13/10	72.5	29.0	20.5	16.2	11.4	9.4	7.5	6.0	4.1
Core #9 50–53	5/13/10	99.8	98.7	94.9	78.9	51.3	36.1	26.9	20.9	14.9
Core #10 0–1	5/13/10	60.5	55.4	49.4	45.4	35.7	27.3	21.9	17.5	15.8
Core #10 20–23	5/13/10	9.9	0.9	0.4	--	--	--	--	--	--
Core #10 80–83	5/13/10	98.0	97.2	93.5	78.0	51.1	38.2	28.9	23.8	18.4
Core #10 108–110	5/13/10	10.5	2.9	0.5	--	--	--	--	--	--
Core #11 0–3	5/18/10	12.7	3.8	1.8	--	--	--	--	--	--
Core #12 0–2	5/13/10	37.5	27.4	20.7	18.9	14.3	10.6	8.2	6.3	5.0
Core #12 29–32	5/13/10	13.0	1.7	0.3	--	--	--	--	--	--
Core #13 0–4	5/18/10	15.0	11.1	7.4	6.7	5.0	4.0	3.1	2.5	2.5
Core #14 0–1	5/18/10	50.5	38.7	29.8	28.0	20.8	14.7	10.9	8.6	6.7
Core #14 2–5	5/18/10	5.1	0.8	0.4	--	--	--	--	--	--
Core #14 54–56	5/18/10	83.0	70.9	66.9	62.9	50.2	35.3	26.5	20.4	17.1
Core #15 2–5	5/18/10	11.5	1.1	0.3	--	--	--	--	--	--
Core #15 17–20	5/18/10	8.5	2.1	0.6	--	--	--	--	--	--
Core #15 22–25	5/18/10	12.2	1.8	0.4	--	--	--	--	--	--

Appendix 2.   Grain-size data from bed-sediment core samples collected in Lower Granite Reservoir and the Clearwater and Snake Rivers just above their confluence, 2010.—Continued

[# , number; cm, centimeter; mm, millimeter; --, not analyzed]

USGS station number	Core identifier	Core sample interval analyzed (cm)	Sample identifier	Date	Percent finer than 16 mm; coarse/ medium pebbles	Percent finer than 8 mm; me- dium/fine pebbles	Percent finer than 4 mm; fine/ very fine pebbles	Percent finer than 2 mm; very fine pebbles/ very coarse sand	Percent finer than 1 mm; very coarse sand/ coarse sand	Percent finer than 0.50 mm; coarse/ medium sand
					Clearwater River—Continued					
462520117004700	15	45–48	Core #15 45–48	5/18/10	--	--	--	100.0	98.4	86.8
462520117004700	15	58–61	Core #15 58–61	5/18/10	--	--	--	100.0	99.9	95.7
462511117002800	16	0–3	Core #16 0–3	5/18/10	--	--	--	--	--	--
462512117002500	17	0–3	Core #17 0–3	5/18/10	--	--	--	--	100.0	90.8
462515117002400	18	0–3	Core #18 0–3	5/18/10	--	--	--	100.0	99.9	90.6
					Lower Granite Reservoir					
463914117245400	19	0–5	Core #19 0–5	4/7/10	--	--	--	--	--	--
463914117245400	19	65–70	Core #19 65–70	4/7/10	--	--	--	--	--	--
463914117245400	19	170–175	Core #19 170–175	4/7/10	--	--	--	--	--	--
463921117245200	20	0–10	Core #20 0–10	4/7/10	--	--	--	--	--	--
463921117245200	20	147–157	Core #20 147–157	4/7/10	--	--	--	--	--	--
463921117245200	20	224–229	Core #20 224–229	4/7/10	--	--	--	--	--	--
463930117244800	21	0–10	Core #21 0–10	4/7/10	--	--	--	--	--	--
463930117244800	21	21–27	Core #21 21–27	4/7/10	--	--	--	--	100.0	99.8
463930117244800	21	30–35	Core #21 30–35	4/7/10	--	--	--	100.0	99.7	98.3
463749117232500	22	0–5	Core #22 0–5	4/7/10	--	--	--	--	--	--
463749117232500	22	92–97	Core #22 92–97	4/7/10	--	--	--	--	--	--
463749117232500	22	158–162	Core #22 158–162	4/7/10	--	--	--	--	--	100.0
463748117231500	23	0–5	Core #23 0–5	4/7/10	--	--	--	--	--	--
463748117231500	23	50–55	Core #23 50–55	4/7/10	--	--	--	--	100.0	99.3
463748117231500	23	161–164	Core #23 161–164	4/7/10	--	--	--	100.0	99.7	95.7
463745117225500	24	sample 1	Core #24 sample 1	4/6/10	100.0	92.8	76.8	73.7	69.7	66.7
463745117225500	24	sample 2	Core #24 sample 2	4/6/10	100.0	61.6	36.7	35.6	33.4	31.7
463516117210100	25	0–5	Core #25 0–5	4/7/10	--	--	--	--	--	--
463516117210100	25	20–25	Core #25 20–25	4/7/10	--	--	--	--	--	--
463516117210100	25	41–46	Core #25 41–46	4/7/10	--	--	--	--	--	--
463519117205200	26	0–5	Core #26 0–5	4/7/10	--	--	--	--	--	--
463519117205200	26	60–65	Core #26 60–65	4/7/10	--	--	--	--	--	--
463519117205200	26	120–125	Core #26 120–125	4/7/10	--	--	--	--	--	--
463521117204600	27	0–5	Core #27 0–5	4/7/10	--	--	--	--	--	--
463521117204600	27	90–95	Core #27 90–95	4/7/10	--	--	--	--	--	--
463521117204600	27	150–155	Core #27 150–155	4/7/10	--	--	--	--	--	--
463521117204600	27	239–244	Core #27 239–244	4/7/10	--	--	--	--	--	--
463316117161800	28	15–20	Core #28 15–20	4/6/10	--	--	--	--	--	--
463316117161800	28	55–60	Core #28 55–60	4/6/10	--	--	--	--	--	--
463316117161800	28	98–102	Core #28 98–102	4/6/10	--	--	--	--	--	--

Appendix 2.   Grain-size data from bed-sediment core samples collected in Lower Granite Reservoir and the Clearwater and Snake Rivers just above their confluence, 2010.—Continued

[# , number; cm, centimeter; mm, millimeter; --, not analyzed]

Sample identifier	Date	Percent finer than 0.25 mm; medium/ fine sand	Percent finer than 0.125 mm; fine/very fine sand	Percent finer than 0.0625 mm; very fine sand/ coarse silt	Percent finer than 0.031 mm; coarse/ medium silt	Percent finer than 0.016 mm; medium/ fine silt	Percent finer than 0.008 mm; fine/very fine silt	Percent finer than 0.004 mm; very fine silt/clay	Percent finer than 0.002 mm; silt/clay for mineral analysis	Percent finer than 0.001 mm; finer than clay
				Clearwater River—Continued						
Core #15 45–48	5/18/10	20.8	2.3	1.0	--	--	--	--	--	--
Core #15 58–61	5/18/10	34.0	13.5	10.4	9.6	7.9	6.2	4.9	3.8	3.2
Core #16 0–3	5/18/10	--	--	97.2	93.3	72.3	59.9	45.4	34.7	21.3
Core #17 0–3	5/18/10	66.6	61.7	59.0	56.8	42.0	31.9	24.5	19.7	17.2
Core #18 0–3	5/18/10	18.3	7.9	4.6	--	--	--	--	--	--
				Lower Granite Reservoir						
Core #19 0–5	4/7/10	--	--	98.8	96.8	80.1	52.9	33.3	22.9	16.7
Core #19 65–70	4/7/10	--	--	99.9	97.1	82.6	54.9	36.3	26.8	17.1
Core #19 170–175	4/7/10	--	--	99.7	95.6	77.1	44.9	26.0	17.1	14.1
Core #20 0–10	4/7/10	--	--	99.7	97.9	83.5	56.4	33.3	22.5	13.8
Core #20 147–157	4/7/10	--	--	99.8	97.8	85.0	52.5	28.3	17.8	10.9
Core #20 224–229	4/7/10	--	--	99.7	97.0	79.1	51.8	31.7	22.5	14.1
Core #21 0–10	4/7/10	--	--	96.4	88.8	66.7	42.9	27.2	18.9	13.1
Core #21 21–27	4/7/10	98.6	89.6	62.3	51.1	37.2	24.5	16.0	11.8	8.5
Core #21 30–35	4/7/10	92.9	80.2	55.0	46.6	35.6	23.7	16.0	11.5	7.5
Core #22 0–5	4/7/10	--	--	99.1	95.9	78.2	49.7	31.2	22.8	16.0
Core #22 92–97	4/7/10	--	--	99.7	97.0	83.1	57.4	36.8	24.8	17.3
Core #22 158–162	4/7/10	99.9	99.8	99.2	94.9	77.3	50.1	31.2	21.3	13.2
Core #23 0–5	4/7/10	--	--	98.4	92.1	67.2	40.8	27.0	20.1	14.1
Core #23 50–55	4/7/10	97.1	96.8	96.1	90.5	67.1	44.3	30.2	22.5	16.6
Core #23 161–164	4/7/10	83.8	78.4	71.8	62.8	47.6	31.0	20.2	14.3	9.2
Core #24 sample 1	4/6/10	63.5	60.1	55.0	44.6	30.0	19.9	13.4	8.9	5.8
Core #24 sample 2	4/6/10	29.7	27.8	25.2	19.6	12.6	7.8	4.9	3.2	2.1
Core #25 0–5	4/7/10	--	--	97.4	90.2	67.4	39.5	25.1	17.7	11.1
Core #25 20–25	4/7/10	--	--	97.8	91.7	65.0	41.7	28.7	21.4	14.2
Core #25 41–46	4/7/10	--	--	98.5	91.6	69.6	43.2	27.8	19.9	12.3
Core #26 0–5	4/7/10	--	--	97.4	87.9	60.2	36.5	23.6	17.7	15.2
Core #26 60–65	4/7/10	--	--	98.5	89.2	64.5	39.2	27.5	19.6	17.4
Core #26 120–125	4/7/10	--	--	98.5	88.7	63.0	38.9	25.1	19.6	14.1
Core #27 0–5	4/7/10	--	--	98.9	94.1	71.0	43.5	28.9	20.3	10.9
Core #27 90–95	4/7/10	--	--	98.8	90.5	63.9	37.5	24.7	17.1	12.0
Core #27 150–155	4/7/10	--	--	97.3	81.4	52.5	30.3	22.3	14.1	5.4
Core #27 239–244	4/7/10	--	--	99.1	92.1	70.0	43.3	28.5	21.5	17.2
Core #28 15–20	4/6/10	--	--	98.1	86.9	57.3	36.9	26.8	18.7	15.0
Core #28 55–60	4/6/10	--	--	99.0	93.0	64.7	43.3	31.5	25.4	20.7
Core #28 98–102	4/6/10	--	--	94.5	76.9	48.0	29.4	21.5	16.8	12.2

Appendix 2.   Grain-size data from bed-sediment core samples collected in Lower Granite Reservoir and the Clearwater and Snake Rivers just above their confluence, 2010.—Continued

[# , number; cm, centimeter; mm, millimeter; --, not analyzed]

USGS station number	Core identifier	Core sample interval analyzed (cm)	Sample identifier	Date	Percent finer than 16 mm; coarse/ medium pebbles	Percent finer than 8 mm; me-dium/fine pebbles	Percent finer than 4 mm; fine/ very fine pebbles	Percent finer than 2 mm; very fine pebbles/ very coarse sand	Percent finer than 1 mm; very coarse sand/ coarse sand	Percent finer than 0.50 mm; coarse/ medium sand
				Lower Granite Reservoir—Continued						
463318117161600	29	10–15	Core #29 10–15	4/6/10	--	--	--	--	--	--
463318117161600	29	70–75	Core #29 70–75	4/6/10	--	--	--	--	--	--
463318117161600	29	87–91	Core #29 87–91	4/6/10	--	--	--	--	--	--
463322117161000	30B	0–3	Core #30B 0–3	4/6/10	--	--	--	--	--	--
463322117161000	30B	10–12	Core #30B 10–12	4/6/10	--	--	--	--	--	--
463216117150001	31	10–14	Core #31 10–14	4/6/10	--	--	--	--	100.0	99.9
463216117150001	31	115–119	Core #31 115–119	4/6/10	--	--	--	--	--	--
463216117150001	31	157–161	Core #31 157–161	4/6/10	--	--	--	--	--	100.0
463219117145900	32	5–12	Core #32 5–12	4/6/10	--	--	--	--	100.0	99.5
463219117145900	32	30–35	Core #32 30–35	4/6/10	--	--	--	--	--	--
463219117145900	32	60–64	Core #32 60–64	4/6/10	--	--	--	--	100.0	99.9
463221117145700	33	3–7	Core #33 3–7	4/6/10	--	--	--	--	--	100.0
463221117145700	33	14–18	Core #33 14–18	4/6/10	--	--	--	--	--	100.0
462655117123600	34	8–11	Core #34 8–11	4/9/10	--	--	--	--	100.0	99.8
462655117123600	34	66–69	Core #34 66–69	4/9/10	--	--	--	--	100.0	99.9
462655117123600	34	122–125	Core #34 122–125	4/9/10	--	--	--	--	--	100.0
462655117123600	34	144–147	Core #34 144–147	4/9/10	--	--	--	--	--	--
462655117123600	34	173–176	Core #34 173–176	4/9/10	--	--	--	--	--	100.0
462520117122001	38	7–10	Core #38 7–10	5/12/10	--	--	--	--	100.0	99.9
462520117122001	38	12–15	Core #38 12–15	5/12/10	--	--	--	--	--	100.0
462520117122001	38	60–63	Core #38 60–63	5/12/10	--	--	--	--	100.0	99.8
462520117122001	38	98–101	Core #38 98–101	5/12/10	--	--	--	--	100.0	99.9
462520117122001	38	105–108	Core #38 105–108	5/12/10	--	--	--	--	100.0	99.9
462520117122001	38	155–158	Core #38 155–158	5/12/10	--	--	--	--	--	--
462520117122001	38A	6–8	Core #38A 6–8	4/9/10	--	--	--	100.0	99.9	99.8
462520117122001	38A	34–36	Core #38A 34–36	4/9/10	--	--	--	--	100.0	99.7
462520117122001	38A	41–43	Core #38A 41–43	4/9/10	--	--	--	--	100.0	99.9
462522117121800	39	12–15	Core #39 12–15	5/12/10	--	--	--	--	100.0	99.6
462522117121800	39	20–23	Core #39 20–23	5/12/10	--	--	--	--	100.0	99.9
462522117121800	39	87–90	Core #39 87–90	5/12/10	--	--	--	--	100.0	99.7
462522117121800	39	136–139	Core #39 136–139	5/12/10	--	--	--	--	--	100.0
462526117102600	40	67–70	Core #40 67–70	5/12/10	--	--	--	--	--	100.0
462526117102600	40	116–119	Core #40 116–119	5/12/10	--	--	--	--	--	100.0
462526117102600	40	122–125	Core #40 122–125	5/12/10	--	--	--	--	100.0	99.9
462526117102600	40	140–143	Core #40 140–143	5/12/10	--	--	--	--	100.0	99.9

Appendix 2. Grain-size data from bed-sediment core samples collected in Lower Granite Reservoir and the Clearwater and Snake Rivers just above their confluence, 2010.—Continued

[# , number; cm, centimeter; mm, millimeter; --, not analyzed]

Sample identifier	Date	Percent finer than 0.25 mm; medium/ fine sand	Percent finer than 0.125 mm; fine/very fine sand	Percent finer than 0.0625 mm; very fine sand/ coarse silt	Percent finer than 0.031 mm; coarse/ medium silt	Percent finer than 0.016 mm; medium/ fine silt	Percent finer than 0.008 mm; fine/very fine silt	Percent finer than 0.004 mm; very fine silt/clay	Percent finer than 0.002 mm; silt/clay for mineral analysis	Percent finer than 0.001 mm; finer than clay
				Lower Granite Reservoir—Continued						
Core #29 10–15	4/6/10	--	--	96.4	87.0	62.4	43.0	29.9	22.1	14.9
Core #29 70–75	4/6/10	--	--	97.6	87.9	64.6	42.3	30.5	23.2	19.4
Core #29 87–91	4/6/10	--	--	93.5	77.6	47.9	29.7	19.8	14.8	11.4
Core #30B 0–3	4/6/10	--	--	93.1	79.0	58.8	39.8	26.9	20.3	18.8
Core #30B 10–12	4/6/10	--	--	92.3	86.8	42.6	46.0	28.6	20.2	17.2
Core #31 10–14	4/6/10	99.8	99.4	85.6	62.1	38.3	26.8	20.5	15.7	14.7
Core #31 115–119	4/6/10	--	--	95.3	82.4	56.8	38.9	27.4	21.3	19.2
Core #31 157–161	4/6/10	99.8	99.0	77.8	63.4	44.4	31.8	23.2	18.7	11.9
Core #32 5–12	4/6/10	99.2	97.3	83.1	69.7	50.9	36.8	26.5	20.2	11.1
Core #32 30–35	4/6/10	--	--	96.4	90.2	70.7	49.8	34.6	25.3	22.4
Core #32 60–64	4/6/10	99.4	96.3	61.6	47.2	31.6	21.9	15.9	12.6	8.8
Core #33 3–7	4/6/10	99.7	94.7	56.3	36.9	22.5	15.5	11.5	8.5	5.3
Core #33 14–18	4/6/10	99.8	98.2	76.9	54.1	32.2	22.2	15.9	12.5	9.0
Core #34 8–11	4/9/10	99.2	90.7	70.2	56.9	40.5	28.9	19.9	14.9	11.3
Core #34 66–69	4/9/10	99.2	83.5	38.5	29.8	22.8	18.9	10.1	10.5	9.9
Core #34 122–125	4/9/10	99.9	96.0	75.6	--	--	--	--	--	--
Core #34 144–147	4/9/10	--	--	92.4	81.7	57.6	42.3	31.9	23.0	18.4
Core #34 173–176	4/9/10	99.9	97.8	85.1	63.9	41.3	28.8	20.6	15.5	9.6
Core #38 7–10	5/12/10	98.9	51.6	37.2	30.1	22.4	17.7	13.6	11.0	10.1
Core #38 12–15	5/12/10	99.7	91.9	84.5	75.3	58.6	48.7	38.2	31.2	26.8
Core #38 60–63	5/12/10	96.1	11.6	1.7	--	--	--	--	--	--
Core #38 98–101	5/12/10	97.4	12.1	1.3	--	--	--	--	--	--
Core #38 105–108	5/12/10	99.6	88.9	79.1	63.8	42.5	31.1	22.9	17.8	11.7
Core #38 155–158	5/12/10	100.0	99.2	96.1	87.7	62.3	42.6	31.8	25.1	17.2
Core #38A 6–8	4/9/10	98.5	60.5	39.5	28.6	19.8	14.5	10.6	7.8	4.9
Core #38A 34–36	4/9/10	99.2	79.1	55.2	37.4	24.2	17.9	13.1	9.4	6.0
Core #38A 41–43	4/9/10	99.7	92.8	86.1	76.1	56.6	40.0	28.2	21.4	15.2
Core #39 12–15	5/12/10	85.9	24.3	4.8	--	--	--	--	--	--
Core #39 20–23	5/12/10	99.5	96.4	58.0	40.9	29.4	24.5	18.6	15.3	13.4
Core #39 87–90	5/12/10	89.2	9.2	2.7	--	--	--	--	--	--
Core #39 136–139	5/12/10	99.9	99.2	90.0	63.3	45.7	35.5	25.4	20.2	16.2
Core #40 67–70	5/12/10	99.4	22.9	1.7	--	--	--	--	--	--
Core #40 116–119	5/12/10	99.9	92.3	86.7	78.0	60.0	44.3	32.8	25.3	21.4
Core #40 122–125	5/12/10	98.4	11.8	3.3	--	--	--	--	--	--
Core #40 140–143	5/12/10	99.7	97.3	75.3	59.3	41.4	30.3	23.3	18.4	14.3

Appendix 2.   Grain-size data from bed-sediment core samples collected in Lower Granite Reservoir and the Clearwater and Snake Rivers just above their confluence, 2010.—Continued

[# , number; cm, centimeter; mm, millimeter; --, not analyzed]

USGS station number	Core identifier	Core sample interval analyzed (cm)	Sample identifier	Date	Percent finer than 16 mm; coarse/ medium pebbles	Percent finer than 8 mm; me- dium/fine pebbles	Percent finer than 4 mm; fine/ very fine pebbles	Percent finer than 2 mm; very fine pebbles/ very coarse sand	Percent finer than 1 mm; very coarse sand/ coarse sand	Percent finer than 0.50 mm; coarse/ medium sand
					Lower Granite Reservoir—Continued					
462529117102800	41	6–8	Core #41 6–8	5/12/10	--	--	100.0	99.9	99.5	97.2
462540117081100	43	3–5	Core #43 3–5	5/12/10	--	--	--	100.0	99.9	99.7
462540117081100	43	15–18	Core #43 15–18	5/12/10	--	--	--	--	100.0	99.7
462540117081100	43	53–56	Core #43 53–56	5/12/10	--	--	--	--	100.0	100.0
462540117081100	43	120–123	Core #43 120–123	5/12/10	--	--	--	100.0	99.4	99.0
462542117081100	44	0–4	Core #44 0–4	5/12/10	--	--	--	100.0	99.0	96.6
462514117065400	46	50–53	Core #46 50–53	5/13/10	--	--	--	--	100.0	99.9
462514117065400	46	55–58	Core #46 55–58	5/13/10	--	--	--	--	--	100.0
462514117065400	46	80–83	Core #46 80–83	5/13/10	--	--	--	--	100.0	99.8
462514117065400	46	97–100	Core #46 97–100	5/13/10	--	--	--	--	100.0	99.9
462515117065401	47	5–8	Core #47 5–8	5/13/10	--	--	--	--	100.0	96.9
462515117065401	47	51–54	Core #47 51–54	5/13/10	--	--	--	100.0	99.3	98.3
462515117065401	47	109–112	Core #47 109–112	5/13/10	--	--	--	--	100.0	99.8
462520117065300	48	0–2	Core #48 0–2	5/12/10	--	--	--	100.0	99.7	95.7
462456117052900	49	2–5	Core #49 2–5	5/14/10	--	--	--	--	100.0	99.9
462456117052900	49	30–33	Core #49 30–33	5/14/10	--	--	--	--	--	100.0
462456117052900	49	86–89	Core #49 86–89	5/14/10	--	--	--	--	100.0	99.6
462456117052900	49	98–101	Core #49 98–101	5/14/10	--	--	--	--	--	100.0
462456117052900	49	128–131	Core #49 128–131	5/14/10	--	--	--	--	100.0	99.9
462501117052800	50	0–2	Core #50 0–2	5/14/10	--	--	--	--	100.0	99.2
462501117052800	50	20–23	Core #50 20–23	5/14/10	--	--	--	--	100.0	99.6
462501117052800	50	62–65	Core #50 62–65	5/14/10	--	--	--	--	100.0	99.7
462510117053900	52	0–14	Core #52 0–14	5/14/10	--	--	--	--	--	--
462510117053900	52	16–19	Core #52 16–19	5/14/10	--	--	--	--	--	--
462510117053900	52	54–57	Core #52 54–57	5/14/10	--	--	--	--	--	--
462510117053900	52	104–107	Core #52 104–107	5/14/10	--	--	--	--	--	--
462506117050400	53	0–2	Core #53 0–2	5/12/10	--	--	--	100.0	99.9	99.5
462506117050400	53	27–30	Core #53 27–30	5/12/10	--	--	--	--	100.0	99.7
462506117050400	53	63–66	Core #53 63–66	5/12/10	--	--	--	--	100.0	98.3
462509117050300	54	2–5	Core #54 2–5	5/15/10	--	--	--	100.0	99.7	94.7
462509117050300	54	73–75	Core #54 73–75	5/15/10	--	--	--	--	100.0	89.7
462509117050300	54	103–106	Core #54 103–106	5/15/10	--	--	--	100.0	99.6	94.7
462520117042900	56	0–5	Core #56 0–5	5/15/10	--	--	--	--	100.0	99.7
462523117042900	57	0–1	Core #57 0–1	5/15/10	--	--	--	--	100.0	99.7

Appendix 2.   Grain-size data from bed-sediment core samples collected in Lower Granite Reservoir and the Clearwater and Snake Rivers just above their confluence, 2010.—Continued

[# , number; cm, centimeter; mm, millimeter; --, not analyzed]

Sample identifier	Date	Percent finer than 0.25 mm; medium/ fine sand	Percent finer than 0.125 mm; fine/very fine sand	Percent finer than 0.0625 mm; very fine sand/ coarse silt	Percent finer than 0.031 mm; coarse/ medium silt	Percent finer than 0.016 mm; medium/ fine silt	Percent finer than 0.008 mm; fine/very fine silt	Percent finer than 0.004 mm; very fine silt/clay	Percent finer than 0.002 mm; silt/clay for mineral analysis	Percent finer than 0.001 mm; finer than clay
Lower Granite Reservoir—Continued										
Core #41 6–8	5/12/10	51.8	2.4	0.7	--	--	--	--	--	--
Core #43 3–5	5/12/10	99.4	97.6	93.6	88.3	70.5	57.5	46.6	37.8	23.9
Core #43 15–18	5/12/10	82.6	4.7	1.0	--	--	--	--	--	--
Core #43 53–56	5/12/10	99.7	96.9	73.3	53.6	34.9	25.6	20.1	16.3	14.4
Core #43 120–123	5/12/10	93.9	60.1	46.4	39.8	32.2	25.5	20.5	16.8	14.9
Core #44 0–4	5/12/10	22.6	3.0	1.1	--	--	--	--	--	--
Core #46 50–53	5/13/10	97.2	16.6	6.0	--	--	--	--	--	--
Core #46 55–58	5/13/10	99.5	83.8	72.0	59.8	43.1	31.8	24.1	18.9	16.2
Core #46 80–83	5/13/10	98.4	46.8	32.2	28.6	23.0	18.6	14.4	11.5	8.2
Core #46 97–100	5/13/10	96.4	11.1	1.7	--	--	--	--	--	--
Core #47 5–8	5/13/10	88.9	12.2	3.6	--	--	--	--	--	--
Core #47 51–54	5/13/10	93.2	13.8	2.3	--	--	--	--	--	--
Core #47 109–112	5/13/10	81.4	9.2	1.7	--	--	--	--	--	--
Core #48 0–2	5/12/10	13.6	1.8	0.6	--	--	--	--	--	--
Core #49 2–5	5/14/10	98.2	7.1	0.8	--	--	--	--	--	--
Core #49 30–33	5/14/10	99.8	99.0	88.1	67.5	45.5	31.5	24.7	19.7	17.9
Core #49 86–89	5/14/10	98.7	80.0	22.4	14.2	10.3	8.0	5.9	4.8	3.4
Core #49 98–101	5/14/10	99.8	96.5	89.9	79.9	60.0	41.6	29.4	21.9	18.8
Core #49 128–131	5/14/10	99.2	60.1	12.6	--	--	--	--	--	--
Core #50 0–2	5/14/10	69.5	36.8	27.7	25.0	21.4	17.8	14.3	11.3	10.0
Core #50 20–23	5/14/10	46.1	1.4	0.4	--	--	--	--	--	--
Core #50 62–65	5/14/10	37.7	2.0	0.6	--	--	--	--	--	--
Core #52 0–14	5/14/10	--	--	96.5	81.8	55.9	39.2	28.6	21.1	15.8
Core #52 16–19	5/14/10	--	--	96.7	81.2	54.4	42.0	32.9	27.2	19.2
Core #52 54–57	5/14/10	--	--	96.5	78.2	55.7	40.9	32.0	26.8	16.8
Core #52 104–107	5/14/10	--	--	98.2	85.0	59.3	41.9	32.2	25.8	19.6
Core #53 0–2	5/12/10	84.8	52.4	39.1	33.6	26.2	20.4	15.9	12.2	7.0
Core #53 27–30	5/12/10	67.8	2.5	0.5	--	--	--	--	--	--
Core #53 63–66	5/12/10	49.0	5.7	2.8	--	--	--	--	--	--
Core #54 2–5	5/15/10	10.0	0.8	0.2	--	--	--	--	--	--
Core #54 73–75	5/15/10	29.5	23.3	16.3	14.4	12.0	9.4	7.5	6.1	3.8
Core #54 103–106	5/15/10	29.7	25.8	20.7	15.8	10.5	7.7	6.0	4.8	3.6
Core #56 0–5	5/15/10	78.7	42.4	34.0	32.7	26.5	20.7	15.9	12.1	7.1
Core #57 0–1	5/15/10	15.0	0.8	0.3	--	--	--	--	--	--

Appendix 2. Grain-size data from bed-sediment core samples collected in Lower Granite Reservoir and the Clearwater and Snake Rivers just above their confluence, 2010.—Continued

[# , number; cm, centimeter; mm, millimeter; --, not analyzed]

USGS station number	Core identifier	Core sample interval analyzed (cm)	Sample identifier	Date	Percent finer than 16 mm; coarse/ medium pebbles	Percent finer than 8 mm; me- dium/fine pebbles	Percent finer than 4 mm; fine/ very fine pebbles	Percent finer than 2 mm; very fine pebbles/ very coarse sand	Percent finer than 1 mm; very coarse sand/ coarse sand	Percent finer than 0.50 mm; coarse/ medium sand
Lower Granite Reservoir—Continued										
462540117033500	59	0–3	Core #59 0–3	5/15/10	--	--	--	--	100.0	99.7
462542117033601	60	0–2	Core #60 0–2	5/15/10	--	--	--	--	100.0	93.3
462542117033601	60	6–9	Core #60 6–9	5/15/10	--	--	--	100.0	99.9	99.1
462542117033601	60	50–53	Core #60 50–53	5/15/10	--	--	--	100.0	99.9	99.3
462541117030300	62	84–87	Core #62 84–87	5/12/10	--	--	--	100.0	99.9	99.6
462541117030300	62	94–97	Core #62 94–97	5/12/10	--	--	--	100.0	99.9	99.2
462546117030500	63	0–2	Core #63 0–2	5/20/10	--	--	--	100.0	98.3	92.3
462546117030500	63	10–13	Core #63 10–13	5/20/10	--	--	--	100.0	99.7	83.7
462536117023500	65	60–63	Core #65 60–63	5/11/10	--	--	--	--	100.0	99.2
462536117023500	65	65–68	Core #65 65–68	5/11/10	--	--	--	--	100.0	99.4
462536117023500	65	85–88	Core #65 85–88	5/11/10	--	--	--	--	100.0	98.3
462542117024801	66	0–1	Core #66 0–1	5/20/10	--	--	--	--	100.0	99.7
462542117024801	66	10–13	Core #66 10–13	5/20/10	--	--	--	100.0	99.9	97.7
Confluence										
462535117020701	70	0–2	Core #70 0–2	5/14/10	--	--	--	--	100.0	99.4
462535117020701	70	4–7	Core #70 4–7	5/14/10	--	--	--	100.0	99.9	97.3
462535117020701	70	26–28	Core #70 26–28	5/14/10	--	--	--	--	100.0	98.7
462535117020701	70	35–36	Core #70 35–36	5/14/10	--	--	--	--	100.0	98.4
462535117020701	70	38–41	Core #70 38–41	5/14/10	--	--	--	100.0	99.8	98.9
462535117020701	70	50–52	Core #70 50–52	5/14/10	--	--	100.0	99.5	98.9	96.0
462537117020500	71	2–5	Core #71 2–5	5/14/10	--	--	--	100.0	99.9	99.3
462537117020500	71	43–46	Core #71 43–46	5/14/10	--	--	--	--	100.0	99.8
462537117020500	71	90–93	Core #71 90–93	5/14/10	--	--	--	--	--	--
462537117020500	71	135–137	Core #71 135–137	5/14/10	--	--	--	--	100.0	99.4
462537117020500	71	144–146	Core #71 144–146	5/14/10	--	--	100.0	99.3	99.1	98.7
Snake River										
462522117021500	72	0–0.5	Core #72 0–0.5	5/19/10	--	--	--	100.0	99.7	97.6
462522117021501	73	0–2	Core #73 0–2	5/19/10	--	--	--	--	100.0	98.4
462440117021300	78	0–0.5	Core #78 0–0.5	5/19/10	--	--	--	--	100.0	99.7
462437117020601	80	0–1	Core #80 0–1	5/19/10	--	--	--	--	100.0	98.3
462426117021901	81	0–3	Core #81 0–3	5/19/10	--	--	--	100.0	99.2	96.3
462426117021600	82	0–3	Core #82 0–3	10/13/10	--	100.0	99.8	99.7	99.2	81.5
462413117022100	84	5–15	Core #84 5–15	10/13/10	--	--	--	100.0	99.9	90.6
462413117022100	84	25–35	Core #84 25–35	10/13/10	--	100.0	99.8	99.7	99.4	89.5

Appendix 2.   Grain-size data from bed-sediment core samples collected in Lower Granite Reservoir and the Clearwater and Snake Rivers just above their confluence, 2010.—Continued

[# , number; cm, centimeter; mm, millimeter; --, not analyzed]

Sample identifier	Date	Percent finer than 0.25 mm; medium/ fine sand	Percent finer than 0.125 mm; fine/very fine sand	Percent finer than 0.0625 mm; very fine sand/ coarse silt	Percent finer than 0.031 mm; coarse/ medium silt	Percent finer than 0.016 mm; medium/ fine silt	Percent finer than 0.008 mm; fine/very fine silt	Percent finer than 0.004 mm; very fine silt/clay	Percent finer than 0.002 mm; silt/clay for mineral analysis	Percent finer than 0.001 mm; finer than clay
				Lower Granite Reservoir—Continued						
Core #59 0–3	5/15/10	55.4	18.1	3.9	2.9	2.2	1.8	1.5	1.2	0.9
Core #60 0–2	5/15/10	80.4	67.0	36.7	33.0	27.1	22.2	18.4	15.5	10.9
Core #60 6–9	5/15/10	22.8	2.3	0.3	--	--	--	--	--	--
Core #60 50–53	5/15/10	23.2	1.5	0.4	--	--	--	--	--	--
Core #62 84–87	5/12/10	50.6	2.3	0.3	--	--	--	--	--	--
Core #62 94–97	5/12/10	74.0	3.5	1.0	--	--	--	--	--	--
Core #63 0–2	5/20/10	26.0	7.6	2.8	--	--	--	--	--	--
Core #63 10–13	5/20/10	5.9	0.7	0.1	--	--	--	--	--	--
Core #65 60–63	5/11/10	26.0	1.1	0.3	--	--	--	--	--	--
Core #65 65–68	5/11/10	43.6	1.6	0.3	--	--	--	--	--	--
Core #65 85–88	5/11/10	14.2	1.1	0.4	--	--	--	--	--	--
Core #66 0–1	5/20/10	96.7	90.8	71.7	58.4	44.6	34.8	28.0	24.2	21.7
Core #66 10–13	5/20/10	14.0	2.0	0.4	--	--	--	--	--	--
				Confluence						
Core #70 0–2	5/14/10	84.5	66.7	58.1	52.0	41.5	33.4	27.4	23.2	21.4
Core #70 4–7	5/14/10	44.1	3.6	1.0	--	--	--	--	--	--
Core #70 26–28	5/14/10	59.6	3.4	0.9	--	--	--	--	--	--
Core #70 35–36	5/14/10	74.7	31.9	28.1	26.5	23.6	20.0	16.5	14.0	12.8
Core #70 38–41	5/14/10	77.2	6.6	2.2	--	--	--	--	--	--
Core #70 50–52	5/14/10	45.0	6.8	2.6	--	--	--	--	--	--
Core #71 2–5	5/14/10	69.9	3.8	1.2	--	--	--	--	--	--
Core #71 43–46	5/14/10	98.1	89.6	81.6	67.1	45.7	34.6	28.0	23.8	19.6
Core #71 90–93	5/14/10	--	--	95.1	76.0	49.1	36.9	29.9	25.2	20.5
Core #71 135–137	5/14/10	94.3	20.0	5.2	4.6	3.7	2.4	1.6	1.3	0.9
Core #71 144–146	5/14/10	97.9	93.5	71.1	52.9	36.2	27.9	21.8	17.4	14.2
				Snake River						
Core #72 0–0.5	5/19/10	16.2	1.0	0.1	--	--	--	--	--	--
Core #73 0–2	5/19/10	26.8	3.6	0.3	--	--	--	--	--	--
Core #78 0–0.5	5/19/10	48.3	3.1	0.2	--	--	--	--	--	--
Core #80 0–1	5/19/10	41.9	6.3	0.8	--	--	--	--	--	--
Core #81 0–3	5/19/10	26.5	1.5	0.2	--	--	--	--	--	--
Core #82 0–3	10/13/10	8.6	0.8	0.3	--	--	--	--	--	--
Core #84 5–15	10/13/10	8.5	1.0	0.4	--	--	--	--	--	--
Core #84 25–35	10/13/10	7.9	0.6	0.3	--	--	--	--	--	--

Appendix 2.   Grain-size data from bed-sediment core samples collected in Lower Granite Reservoir and the Clearwater and Snake Rivers just above their confluence, 2010.—Continued

[# , number; cm, centimeter; mm, millimeter; --, not analyzed]

USGS station number	Core identifier	Core sample interval analyzed (cm)	Sample identifier	Date	Percent finer than 16 mm; coarse/ medium pebbles	Percent finer than 8 mm; me- dium/fine pebbles	Percent finer than 4 mm; fine/ very fine pebbles	Percent finer than 2 mm; very fine pebbles/ very coarse sand	Percent finer than 1 mm; very coarse sand/ coarse sand	Percent finer than 0.50 mm; coarse/ medium sand
Snake River—Continued										
462413117022100	84	45–55	Core #84 45–55	10/13/10	--	--	--	100.0	99.7	91.2
462413117022100	84	65–75	Core #84 65–75	10/13/10	--	--	100.0	99.6	96.8	70.0
462413117022100	84	85–95	Core #84 85–95	10/13/10	--	--	--	100.0	99.9	90.0
462413117021900	85	5–15	Core #85 5–15	10/13/10	--	--	--	100.0	99.6	82.6
462413117021900	85	25–35	Core #85 25–35	10/13/10	--	100.0	99.9	99.5	83.1	6.2
462413117021900	85	45–55	Core #85 45–55	10/13/10	--	--	--	100.0	98.3	67.2
462413117021900	85	55–65	Core #85 55–65	10/13/10	--	--	100.0	99.7	98.8	79.3
462348117022801	87	0–19	Core #87 0–19	5/19/10	--	--	--	100.0	97.9	75.1
462348117022801	87	16–20	Core #87 16–20	10/13/10	--	100.0	99.8	99.8	99.8	96.3
462348117022801	87	50–55	Core #87 50–55	10/13/10	--	--	--	100.0	99.6	94.0
462348117022801	87	68–72	Core #87 68–72	10/13/10	--	--	--	--	--	--
462348117022801	87	77–82	Core #87 77–82	10/13/10	--	--	--	--	100.0	99.5
462352117022101	88	0–3	Core #88 0–3	5/19/10	--	--	--	100.0	99.9	97.8
462348117021701	89	0–3	Core #89 0–3	5/19/10	--	--	--	100.0	99.9	99.0

Appendix 2. Grain-size data from bed-sediment core samples collected in Lower Granite Reservoir and the Clearwater and Snake Rivers just above their confluence, 2010.—Continued

[# , number; cm, centimeter; mm, millimeter; --, not analyzed]

Sample identifier	Date	Percent finer than 0.25 mm; medium/ fine sand	Percent finer than 0.125 mm; fine/very fine sand	Percent finer than 0.0625 mm; very fine sand/ coarse silt	Percent finer than 0.031 mm; coarse/ medium silt	Percent finer than 0.016 mm; medium/ fine silt	Percent finer than 0.008 mm; fine/very fine silt	Percent finer than 0.004 mm; very fine silt/clay	Percent finer than 0.002 mm; silt/clay for mineral analysis	Percent finer than 0.001 mm; finer than clay
				Snake River—Continued						
Core #84 45–55	10/13/10	6.8	0.7	0.3	--	--	--	--	--	--
Core #84 65–75	10/13/10	4.0	0.7	0.4	--	--	--	--	--	--
Core #84 85–95	10/13/10	5.9	0.7	0.3	--	--	--	--	--	--
Core #85 5–15	10/13/10	2.8	0.3	0.2	--	--	--	--	--	--
Core #85 25–35	10/13/10	0.7	0.2	0.2	--	--	--	--	--	--
Core #85 45–55	10/13/10	1.7	0.3	0.2	--	--	--	--	--	--
Core #85 55–65	10/13/10	6.2	0.8	0.2	--	--	--	--	--	--
Core #87 0–19	5/19/10	13.4	4.3	0.6	--	--	--	--	--	--
Core #87 16–20	10/13/10	17.3	0.8	0.2	--	--	--	--	--	--
Core #87 50–55	10/13/10	53.5	40.5	35.4	32.5	24.4	16.9	12.3	9.8	8.1
Core #87 68–72	10/13/10	--	100.0	94.0	85.0	62.2	43.5	32.3	25.5	17.7
Core #87 77–82	10/13/10	91.2	52.4	21.7	18.4	14.8	12.6	10.7	9.0	7.6
Core #88 0–3	5/19/10	73.7	7.9	0.5	--	--	--	--	--	--
Core #89 0–3	5/19/10	57.8	13.6	1.1	--	--	--	--	--	--

Appendix 3—Analytical Data for Major and Trace Elements in Bed-Sediment Core Samples

Appendix 3.   Analytical data for major and trace elements in bed-sediment core samples collected in Lower Granite Reservoir and the Clearwater and Snake Rivers just above their confluence, 2010.

[Concentrations in micrograms per gram unless otherwise noted; #, number; <, less than lower reporting limit]

USGS station number	Sample identifier	Date	Set number	Calcium	Magne-sium	Potas-sium	Sodium	Total carbon (percent)	Inorganic carbon (percent)	Phos-phorus	Alumi-num
				Clearwater River							
462532117011101	Core #9 18–21	5/13/2010	11150	21,800	10,600	13,300	16,100	1.92	0.02	950	80,400
462532117011101	Core #9 34–37	5/13/2010	11150	24,200	10,700	15,100	20,600	1.02	0.01	818	74,500
462532117011101	Core #9 50–53	5/13/2010	11150	20,200	9,920	13,400	15,600	2.84	0.01	956	80,100
				Lower Granite Reservoir							
463216117150001	Core #31 0–3	4/6/2010	11150	22,400	10,200	16,500	15,700	6.02	0.04	1,090	67,900
463216117150001	Core #31 14–17	4/6/2010	11150	22,600	10,700	19,700	18,900	3.76	0.03	1,020	73,900
463216117150001	Core #31 28–31	4/6/2010	11150	21,400	11,100	19,000	14,400	6.55	0.06	1,230	74,200
463216117150001	Core #31 42–45	4/6/2010	11150	22,900	11,300	16,300	13,700	7.62	0.05	1,320	68,300
463216117150001	Core #31 56–59	4/6/2010	11150	23,700	11,700	14,600	13,700	7.90	0.06	1,420	66,300
463216117150001	Core #31 70–73	4/6/2010	11150	22,800	11,900	18,900	16,000	5.07	0.05	1,060	69,300
463216117150001	Core #31 84–87	4/6/2010	11150	24,100	12,300	18,000	14,900	6.82	0.09	1,260	70,700
463216117150001	Core #31 98–101	4/6/2010	11150	25,700	12,000	17,900	18,200	4.44	0.04	1,060	72,300
463216117150001	Core #31 112–115	4/6/2010	11150	22,900	11,500	17,100	15,900	5.00	0.04	1,120	70,700
463216117150001	Core #31 126–129	4/6/2010	11150	25,300	12,600	16,900	16,300	4.88	0.05	1,180	72,500
463216117150001	Core #31 140–143	4/6/2010	11150	22,600	11,400	16,400	14,100	6.28	0.05	1,250	70,500
463216117150001	Core #31 154–157	4/6/2010	11150	25,400	11,400	16,500	17,100	4.15	0.03	1,090	72,400
463216117150001	Core #31 164–167	4/6/2010	11150	27,800	12,700	16,600	18,800	2.78	0.04	1,020	74,800
462520117122001	Core #38 7–10	5/12/2010	11150	26,200	10,700	18,500	20,100	2.04	0.01	864	70,400
462520117122001	Core #38 12–15	5/12/2010	11150	22,600	10,200	13,800	13,600	6.11	0.04	1,340	64,900
462520117122001	Core #38 60–63	5/12/2010	11150	21,900	10,600	20,100	23,500	0.22	0.01	611	74,300
462520117122001	Core #38 98–101	5/12/2010	11150	25,000	11,300	19,500	24,100	0.17	0.02	722	76,200
462520117122001	Core #38 105–108	5/12/2010	11150	30,500	12,300	12,800	17,100	3.29	0.07	1,070	72,900
462520117122001	Core #38 155–158	5/12/2010	11150	25,900	11,000	12,300	14,700	3.65	0.07	1,180	71,700
462515117065401	Core #47 5–8	5/13/2010	11150	18,900	8,340	22,500	22,600	4.43	0.02	561	72,100
462515117065401	Core #47 51–54	5/13/2010	11150	18,900	9,030	23,000	24,500	0.94	0.02	529	78,200
462515117065401	Core #47 109–112	5/13/2010	11150	22,600	9,880	19,900	24,800	0.16	0.01	602	74,600
462542117033601	Core #60 0–2	5/15/2010	11150	18,600	7,280	13,400	11,400	17.10	0.06	831	45,500
462542117033601	Core #60 6–9	5/15/2010	11150	18,600	6,120	23,300	22,900	0.12	<0.01	432	66,400
462542117033601	Core #60 50–53	5/15/2010	11150	26,300	9,640	18,900	21,900	0.12	0.01	521	65,600
462542117024801	Core #66 10–13	5/20/2010	11150	19,600	6,540	21,300	22,400	0.14	<0.01	413	64,900

Appendix 3. Analytical data for major and trace elements in bed-sediment core samples collected in Lower Granite Reservoir and the Clearwater and Snake Rivers just above their confluence, 2010.—Continued

[Concentrations in micrograms per gram unless otherwise noted; #, number; <, less than lower reporting limit]

USGS station number	Sample identifier	Date	Set number	Calcium	Magne- sium	Potas- sium	Sodium	Total carbon (percent)	Inorganic carbon (percent)	Phos- phorus	Alumi- num
				Confluence							
462535117020701	Core #70 0–2	5/14/2010	11150	17,400	8,470	14,900	15,900	5.69	0.03	1,040	71,300
462535117020701	Core #70 4–7	5/14/2010	11150	18,900	6,810	20,400	23,700	0.21	<0.01	420	71,700
462535117020701	Core #70 26–28	5/14/2010	11150	21,400	8,260	19,300	23,400	0.18	<0.01	474	73,200
462535117020701	Core #70 35–36	5/14/2010	11150	17,300	8,270	13,700	14,900	5.64	0.04	1,080	68,600
462535117020701	Core #70 38–41	5/14/2010	11150	22,200	8,960	19,200	23,900	0.42	<0.01	544	76,900
462535117020701	Core #70 50–52	5/14/2010	11150	19,100	8,130	20,000	22,900	0.74	<0.01	514	75,600
				Snake River							
462522117021501	Core #73 0–2	5/19/2010	11150	19,500	7,410	20,700	23,500	0.14	<0.01	440	69,000
462437117020601	Core #80 0–1	5/19/2010	11150	20,500	8,720	20,800	23,700	0.26	0.01	529	70,500
462426117021901	Core #81 0–3	5/19/2010	11150	17,500	6,840	23,200	23,000	0.13	<0.01	384	70,700
462413117022101	Core #84 5–15	10/13/2010	11637	20,200	7,600	19,300	21,000	0.11	<0.01	471	61,200
462413117022101	Core #84 25–35	10/13/2010	11637	18,700	7,140	18,600	20,100	0.11	<0.01	468	59,300
462413117022101	Core #84 45–55	10/13/2010	11637	19,400	7,060	21,000	22,100	0.10	<0.01	453	65,500
462413117022101	Core #84 65–75	10/13/2010	11637	15,000	6,690	21,900	19,500	0.12	0.01	400	60,300
462413117022101	Core #84 85–95	10/13/2010	11637	15,900	5,890	19,500	19,600	0.11	0.01	402	57,700
462412117021801	Core #85 5–15	10/13/2010	11637	17,500	6,120	19,900	20,000	0.17	0.01	397	58,200
462412117021801	Core #85 25–35	10/13/2010	11637	18,300	6,740	18,000	19,400	0.12	<0.01	438	56,400
462412117021801	Core #85 45–55	10/13/2010	11637	21,800	8,770	18,600	20,500	0.12	<0.01	506	61,300
462412117021801	Core #85 55–65	10/13/2010	11637	19,300	7,180	18,600	20,000	0.11	<0.01	477	58,800
462348117022801	Core #87 0–19	5/19/2010	11150	18,500	7,330	19,400	21,400	0.59	0.01	472	63,700
462352117022101	Core #88 0–3	5/19/2010	11150	19,400	7,770	23,500	24,600	0.16	0.01	520	75,200
462348117021701	Core #89 0–3	5/19/2010	11150	19,200	7,640	23,900	24,000	0.17	0.01	499	72,800

Appendix 3.   Analytical data for major and trace elements in bed-sediment core samples collected in Lower Granite Reservoir and the Clearwater and Snake Rivers just above their confluence, 2010.—Continued

[Concentrations in micrograms per gram unless otherwise noted; #, number; <, less than lower reporting limit]

Sample identifier	Date	Barium	Beryl-lium	Bis-muth	Cad-mium	Cerium	Cesium	Chro-mium	Cobalt	Copper	Gallium	Iron
						Clearwater River						
Core #9 18–21	5/13/2010	684	1.7	5.48	0.18	64.4	2.7	46.7	25.2	33.5	20.0	54,000
Core #9 34–37	5/13/2010	735	1.2	4.78	0.12	48.0	1.8	35.4	17.4	740	17.3	40,600
Core #9 50–53	5/13/2010	676	1.5	5.63	0.21	67.3	3.1	44.8	23.3	38.2	19.9	52,800
						Lower Granite Reservoir						
Core #31 0–3	4/6/2010	673	1.8	6.70	0.34	60.9	4.2	42.1	18.1	33.1	19.4	40,900
Core #31 14–17	4/6/2010	765	2.0	6.80	0.29	68.2	4.1	48.7	17.3	30.2	20.1	38,400
Core #31 28–31	4/6/2010	689	2.6	9.61	0.41	91.1	6.8	46.4	23.7	46.9	22.5	43,500
Core #31 42–45	4/6/2010	664	2.1	8.50	0.46	75.4	5.2	49.1	22.2	43.4	21.0	44,500
Core #31 56–59	4/6/2010	644	2.1	7.70	0.42	71.1	4.6	44.9	23.2	46.1	19.9	47,100
Core #31 70–73	4/6/2010	775	2.1	14.3	0.37	76.8	5.3	52.9	18.9	38.6	20.6	40,800
Core #31 84–87	4/6/2010	684	2.2	9.01	0.45	83.3	5.6	53.6	21.4	44.5	21.2	44,900
Core #31 98–101	4/6/2010	714	1.7	6.49	0.32	65.5	4.3	52.9	21.2	36.7	19.8	42,000
Core #31 112–115	4/6/2010	667	1.8	7.08	0.41	69.9	4.7	51.8	19.6	39.9	20.1	42,300
Core #31 126–129	4/6/2010	688	1.9	7.38	0.37	65.7	4.6	54.1	22.8	44.8	20.8	45,600
Core #31 140–143	4/6/2010	655	2.2	8.81	0.42	74.3	5.4	51.0	22.8	44.6	20.7	44,400
Core #31 154–157	4/6/2010	697	1.7	6.64	0.36	59.2	3.9	51.7	22.4	36.2	19.7	42,900
Core #31 164–167	4/6/2010	723	1.5	5.90	0.27	58.1	3.2	52.4	20.9	36.7	19.8	45,000
Core #38 7–10	5/12/2010	762	1.6	5.21	0.19	60.3	2.7	48.1	16.9	24.5	18.7	37,400
Core #38 12–15	5/12/2010	619	1.8	6.41	0.34	56.4	4.0	44.9	20.0	35.0	18.5	41,800
Core #38 60–63	5/12/2010	822	1.6	4.62	0.10	46.8	2.2	43.8	15.7	18.5	17.3	32,400
Core #38 98–101	5/12/2010	804	1.7	4.56	0.11	56.6	2.1	49.0	16.3	18.2	17.7	33,400
Core #38 105–108	5/12/2010	674	1.4	4.66	0.24	52.9	2.5	38.5	27.6	42.5	20.1	63,400
Core #38 155–158	5/12/2010	679	1.4	5.10	0.25	63.7	2.8	42.4	26.4	39.2	20.3	58,600
Core #47 5–8	5/13/2010	868	2.0	5.02	0.17	41.2	2.7	34.3	12.8	17.1	17.2	26,800
Core #47 51–54	5/13/2010	869	1.7	4.82	0.09	43.2	2.8	35.7	12.1	17.0	17.6	26,800
Core #47 109–112	5/13/2010	824	1.6	4.51	0.09	39.6	2.0	40.6	13.5	16.0	16.9	29,500
Core #60 0–2	5/15/2010	547	1.7	5.07	0.32	43.1	2.9	31.6	13.6	41.2	13.3	27,200
Core #60 6–9	5/15/2010	886	1.5	4.57	0.06	31.7	1.9	27.2	9.9	13.0	15.8	19,900
Core #60 50–53	5/15/2010	771	1.3	4.17	0.09	62.0	1.5	56.4	13.6	16.7	16.5	34,000
Core #66 10–13	5/20/2010	838	1.4	4.38	0.06	38.5	1.6	28.0	10.0	13.4	15.5	21,200

Appendix 3.   Analytical data for major and trace elements in bed-sediment core samples collected in Lower Granite Reservoir and the Clearwater and Snake Rivers just above their confluence, 2010.—Continued

[Concentrations in micrograms per gram unless otherwise noted; #, number; <, less than lower reporting limit]

Sample identifier	Date	Barium	Beryl-lium	Bis-muth	Cad-mium	Cerium	Cesium	Chro-mium	Cobalt	Copper	Gallium	Iron
						Confluence						
Core #70 0–2	5/14/2010	700	1.5	5.60	0.18	48.5	2.9	40.4	14.6	39.6	17.4	37,800
Core #70 4–7	5/14/2010	891	1.3	5.08	0.06	36.7	1.5	23.3	8.7	10.7	15.8	22,400
Core #70 26–28	5/14/2010	853	1.4	4.94	0.06	43.4	1.5	28.8	9.9	12.3	16.4	27,100
Core #70 35–36	5/14/2010	666	1.5	5.27	0.20	50.7	2.4	41.9	16.5	27.4	17.0	36,500
Core #70 38–41	5/14/2010	848	1.4	5.05	0.07	37.3	1.7	32.4	11.8	12.9	17.1	30,100
Core #70 50–52	5/14/2010	886	1.4	4.96	0.08	32.7	1.8	26.5	12.2	33.8	16.7	29,800
						Snake River						
Core #73 0–2	5/19/2010	839	1.6	4.24	0.07	36.3	1.7	31.9	9.5	11.4	15.3	21,300
Core #80 0–1	5/19/2010	828	1.4	4.38	0.09	55.6	1.8	39.9	11.1	12.4	16.2	25,400
Core #81 0–3	5/19/2010	920	1.6	4.64	0.07	51.7	2.0	35.1	11.1	12.6	16.5	27,200
Core #84 5–15	10/13/2010	943	1.5	4.11	0.08	30.9	1.6	28.7	11.1	14.8	15.2	24,300
Core #84 25–35	10/13/2010	910	1.6	4.34	0.06	32.3	1.6	27.1	10.3	15.4	14.5	23,100
Core #84 45–55	10/13/2010	1,040	1.8	4.50	0.07	29.4	1.9	24.6	10.8	14.4	16.0	22,600
Core #84 65–75	10/13/2010	1,040	1.8	10.4	0.02	34.7	2.5	22.9	11.1	18.9	16.3	22,300
Core #84 85–95	10/13/2010	1,000	1.7	4.38	0.06	30.1	1.9	20.3	9.7	74.6	14.5	19,500
Core #85 5–15	10/13/2010	936	1.3	5.06	0.07	31.2	1.7	23.1	9.2	12.4	14.4	20,000
Core #85 25–35	10/13/2010	855	1.4	4.03	0.06	39.2	1.6	20.1	10.0	13.2	13.9	22,700
Core #85 45–55	10/13/2010	923	1.5	4.40	0.09	31.0	1.6	31.0	12.3	17.2	15.2	27,500
Core #85 55–65	10/13/2010	886	1.5	4.02	0.07	35.4	1.6	25.0	10.4	14.9	14.6	23,900
Core #87 0–19	5/19/2010	776	1.3	4.08	0.07	34.8	1.6	26.6	10.8	17.7	14.7	24,000
Core #88 0–3	5/19/2010	879	1.8	4.80	0.08	56.2	2.3	36.2	10.4	12.7	17.4	25,100
Core #89 0–3	5/19/2010	883	1.6	4.81	0.08	43.5	2.3	35.7	10.4	12.4	17.1	24,500

Appendix 3. Analytical data for major and trace elements in bed-sediment core samples collected in Lower Granite Reservoir and the Clearwater and Snake Rivers just above their confluence, 2010.—Continued

[Concentrations in micrograms per gram unless otherwise noted; #, number; <, less than lower reporting limit]

Sample identifier	Date	Lantha-num	Lead	Lithium	Manga-nese	Mercury	Molyb-denum	Nickel	Niobium	Rubidium	Scandium
					Clearwater River						
Core #9 18–21	5/13/2010	31.8	14.5	20.7	970	0.03	0.76	23.4	13	62.9	20.2
Core #9 34–37	5/13/2010	24.5	12.5	14.8	746	0.02	0.55	17.0	10	56.1	15.1
Core #9 50–53	5/13/2010	33.3	14.6	22.2	998	0.03	0.80	22.1	13	65.7	19.9
					Lower Granite Reservoir						
Core #31 0–3	4/6/2010	32.8	17.5	26.8	817	0.08	1.2	20.2	14	80.0	15.3
Core #31 14–17	4/6/2010	36.2	17.4	25.4	735	0.08	1.1	20.9	15	90.4	14.1
Core #31 28–31	4/6/2010	49.4	24.3	33.2	1,090	0.14	1.6	22.0	18	101.0	15.2
Core #31 42–45	4/6/2010	38.9	21.7	32.5	1,160	0.28	1.8	24.0	16	83.9	16.1
Core #31 56–59	4/6/2010	38.2	19.9	27.2	1,110	0.14	1.5	23.8	14	74.5	17.6
Core #31 70–73	4/6/2010	40.6	38.7	27.4	852	0.20	2.1	27.4	17	92.7	14.7
Core #31 84–87	4/6/2010	43.6	23.7	29.2	1,100	0.24	1.9	28.2	18	90.6	16.0
Core #31 98–101	4/6/2010	34.4	16.9	24.2	853	0.08	1.2	25.4	17	85.4	15.4
Core #31 112–115	4/6/2010	36.4	18.4	26.6	787	0.09	1.9	25.9	16	86.2	15.7
Core #31 126–129	4/6/2010	34.9	19.4	26.8	964	0.15	2.8	27.5	16	81.7	16.9
Core #31 140–143	4/6/2010	39.7	22.6	28.7	984	0.14	3.4	25.5	16	84.5	16.2
Core #31 154–157	4/6/2010	31.6	17.5	24.5	783	0.09	2.1	24.0	14	75.6	16.2
Core #31 164–167	4/6/2010	30.0	15.5	20.7	790	0.09	0.86	26.2	14	70.8	17.5
Core #38 7–10	5/12/2010	31.9	13.9	19.1	694	0.05	0.69	19.9	13	75.1	14.6
Core #38 12–15	5/12/2010	30.5	16.8	23.6	930	0.08	1.0	22.5	12	67.5	16.2
Core #38 60–63	5/12/2010	24.7	12.4	17.5	532	0.02	1.3	18.1	12	76.1	12.1
Core #38 98–101	5/12/2010	31.0	12.0	15.9	571	0.02	0.58	17.3	13	71.1	13.5
Core #38 105–108	5/12/2010	25.8	12.3	17.8	1,230	0.04	1.0	21.8	12	57.4	23.5
Core #38 155–158	5/12/2010	31.3	13.6	20.6	1,220	0.04	0.96	22.4	13	61.0	22.9
Core #47 5–8	5/13/2010	22.9	13.5	20.0	482	0.03	0.88	15.1	12	88.6	9.1
Core #47 51–54	5/13/2010	22.9	12.9	21.0	437	0.02	0.47	14.9	13	93.7	9.3
Core #47 109–112	5/13/2010	20.6	12.1	14.9	509	0.04	0.51	16.8	11	72.4	11.6
Core #60 0–2	5/15/2010	24.5	13.5	18.2	798	0.07	1.3	17.2	10	64.0	8.3
Core #60 6–9	5/15/2010	17.0	12.4	13.2	349	0.01	0.42	11.8	9.2	84.7	7.6
Core #60 50–53	5/15/2010	33.4	11.1	11.7	753	0.26	0.64	17.6	16	65.8	13.7
Core #66 10–13	5/20/2010	20.1	11.7	12.1	394	0.02	0.54	12.8	8.8	76.0	8.4

Appendix 3. Analytical data for major and trace elements in bed-sediment core samples collected in Lower Granite Reservoir and the Clearwater and Snake Rivers just above their confluence, 2010.—Continued

[Concentrations in micrograms per gram unless otherwise noted; #, number; <, less than lower reporting limit]

Sample identifier	Date	Lantha-num	Lead	Lithium	Manga-nese	Mercury	Molyb-denum	Nickel	Niobium	Rubidium	Scandium
					Confluence						
Core #70 0–2	5/14/2010	27.2	14.5	21.5	673	0.05	0.63	20.1	10	69.3	13.1
Core #70 4–7	5/14/2010	19.9	13.6	13.1	346	0.01	0.31	10.0	7.0	66.9	8.1
Core #70 26–28	5/14/2010	23.0	13.1	13.0	437	0.01	0.32	11.5	9.5	64.4	9.8
Core #70 35–36	5/14/2010	27.3	14.1	20.0	734	0.06	0.76	21.1	8.7	58.1	13.4
Core #70 38–41	5/14/2010	19.3	13.0	14.1	480	0.01	0.37	13.0	8.5	65.6	11.0
Core #70 50–52	5/14/2010	17.7	12.9	15.0	478	0.01	0.35	12.6	7.9	70.8	9.9
					Snake River						
Core #73 0–2	5/19/2010	18.4	11.7	13.3	411	0.01	0.46	12.7	8.2	72.4	8.8
Core #80 0–1	5/19/2010	30.6	11.7	14.0	553	0.02	0.56	15.2	11	75.7	10.6
Core #81 0–3	5/19/2010	28.6	12.2	15.0	574	2.32	0.50	12.9	15	85.3	8.5
Core #84 5–15	10/13/2010	15.9	11.3	13.7	491	0.01	0.44	12.4	9.7	71.7	9.7
Core #84 25–35	10/13/2010	16.6	11.8	13.4	418	0.01	0.43	11.7	8.5	69.9	8.9
Core #84 45–55	10/13/2010	15.6	12.3	15.6	394	0.03	0.44	11.7	8.8	80.2	8.7
Core #84 65–75	10/13/2010	18.8	28.6	20.8	389	0.02	0.47	12.7	12	92.3	7.6
Core #84 85–95	10/13/2010	15.8	11.8	14.4	346	0.01	0.40	10.0	8.7	79.5	7.2
Core #85 5–15	10/13/2010	16.7	13.7	11.8	368	0.01	0.42	10.6	8.7	76.2	7.5
Core #85 25–35	10/13/2010	21.4	11.1	11.4	410	0.01	0.44	10.0	7.8	68	8.7
Core #85 45–55	10/13/2010	15.8	12.0	12.5	535	0.15	0.50	13.1	8.2	69.1	11.4
Core #85 55–65	10/13/2010	19.0	11.0	12.4	439	0.01	0.53	11.2	8.3	70.3	9.3
Core #87 0–19	5/19/2010	18.7	10.9	12.6	470	0.01	0.49	12.5	8.2	71.3	9.5
Core #88 0–3	5/19/2010	30.4	13.0	17.1	562	0.02	0.45	13.8	12	90.4	9.0
Core #89 0–3	5/19/2010	24.8	12.9	15.7	500	0.02	0.44	14.4	13	90.3	8.9

Appendix 3.   Analytical data for major and trace elements in bed-sediment core samples collected in Lower Granite Reservoir and the Clearwater and Snake Rivers just above their confluence, 2010.—Continued

[Concentrations in micrograms per gram unless otherwise noted; #, number; <, less than lower reporting limit]

Sample identifier	Date	Silver	Stron-tium	Thal-lium	Tita-nium	Vana-dium	Yttrium	Zinc	Anti-mony	Arsenic	Organic carbon, percent	Tho-rium	Uranium
					Clearwater River								
Core #9 18–21	5/13/2010	<0.01	281	0.38	7,760	201	27.6	98.2	0.39	3.2	1.90	7.95	2.29
Core #9 34–37	5/13/2010	<0.01	355	0.31	5,720	146	19.9	390	0.23	2.1	1.01	5.71	1.60
Core #9 50–53	5/13/2010	<0.01	265	0.40	7,280	191	28.7	99.3	0.45	3.8	2.83	8.45	2.46
					Lower Granite Reservoir								
Core #31 0–3	4/6/2010	<0.01	295	0.45	4,990	130	27.8	107	1.0	9.7	5.98	8.71	5.57
Core #31 14–17	4/6/2010	<0.01	333	0.47	4,750	122	24.6	113	1.1	8.8	3.73	9.76	4.65
Core #31 28–31	4/6/2010	0.094	278	0.54	4,630	122	36.0	117	1.5	23.5	6.49	13.2	17.0
Core #31 42–45	4/6/2010	0.099	286	0.49	5,040	135	32.2	119	1.9	18.1	7.57	10.4	15.5
Core #31 56–59	4/6/2010	0.020	283	0.44	5,450	150	33.5	116	1.1	13.3	7.84	9.88	12.9
Core #31 70–73	4/6/2010	0.095	297	0.52	4,620	124	29.4	108	1.7	10.6	5.02	11.6	11.7
Core #31 84–87	4/6/2010	0.037	291	0.50	5,110	134	33.1	117	1.5	11.6	6.73	12.7	14.1
Core #31 98–101	4/6/2010	<0.01	328	0.47	5,210	131	26.9	103	0.81	7.5	4.40	10.7	4.93
Core #31 112–115	4/6/2010	<0.01	295	0.48	5,100	129	29.7	109	0.85	8.6	4.96	11.2	6.11
Core #31 126–129	4/6/2010	<0.01	310	0.46	5,480	147	29.2	111	1.8	11.1	4.83	9.82	11.1
Core #31 140–143	4/6/2010	<0.01	274	0.48	5,180	136	32.4	117	1.2	12.1	6.23	11.2	12.8
Core #31 154–157	4/6/2010	<0.01	318	0.42	5,470	141	26.1	107	0.80	8.7	4.12	8.58	6.04
Core #31 164–167	4/6/2010	<0.01	349	0.40	6,170	160	25.2	102	0.81	6.1	2.74	7.62	3.73
Core #38 7–10	5/12/2010	<0.01	365	0.40	5,150	134	21.7	85.2	0.59	4.2	2.03	8.57	3.22
Core #38 12–15	5/12/2010	0.033	272	0.40	4,980	136	29.3	108	0.77	7.4	6.07	7.83	5.86
Core #38 60–63	5/12/2010	<0.01	383	0.40	4,540	117	15.6	68.9	0.77	2.7	0.21	6.38	1.54
Core #38 98–101	5/12/2010	<0.01	390	0.38	5,000	130	17.5	69.5	0.70	2.6	0.15	7.07	1.56
Core #38 105–108	5/12/2010	<0.01	309	0.36	8,320	241	32.1	113	0.45	3.5	3.22	6.36	2.45
Core #38 155–158	5/12/2010	<0.01	277	0.38	8,030	225	32.6	115	0.46	3.9	3.58	6.96	2.81
Core #47 5–8	5/13/2010	<0.01	369	0.45	3,250	83.1	15.6	63.1	0.66	4.1	4.41	6.06	3.47
Core #47 51–54	5/13/2010	<0.01	381	0.46	3,290	81.8	13.6	61.9	0.46	3.0	0.92	6.42	2.37
Core #47 109–112	5/13/2010	<0.01	396	0.37	4,060	109	14.7	62.8	0.53	2.5	0.15	5.76	1.49
Core #60 0–2	5/15/2010	<0.01	239	0.35	2,560	80.8	20.3	79.5	1.1	9.0	17.04	5.34	8.17
Core #60 6–9	5/15/2010	<0.01	395	0.41	2,320	70.2	10.7	46.1	0.40	2.4	0.12	5.05	1.17
Core #60 50–53	5/15/2010	<0.01	394	0.33	5,720	141	19.2	61.4	0.47	2.7	0.11	7.41	1.59
Core #66 10–13	5/20/2010	<0.01	371	0.39	2,730	78.3	11.5	45.2	0.56	2.5	0.14	5.13	1.29

Appendix 3.   Analytical data for major and trace elements in bed-sediment core samples collected in Lower Granite Reservoir and the Clearwater and Snake Rivers just above their confluence, 2010.—Continued

[Concentrations in micrograms per gram unless otherwise noted; #, number; <, less than lower reporting limit]

Sample identifier	Date	Silver	Stron-tium	Thal-lium	Tita-nium	Vana-dium	Yttrium	Zinc	Anti-mony	Arsenic	Organic carbon, percent	Tho-rium	Uranium
					Confluence								
Core #70 0–2	5/14/2010	<0.01	302	0.39	4,500	112	22.2	92.5	0.38	4.1	5.66	5.72	3.28
Core #70 4–7	5/14/2010	<0.01	414	0.35	2,720	72.1	9.8	47.4	0.10	1.1	0.21	4.96	0.98
Core #70 26–28	5/14/2010	<0.01	423	0.32	3,910	89.7	13.6	54.6	0.20	1.0	0.18	5.34	1.20
Core #70 35–36	5/14/2010	<0.01	277	0.35	4,280	114	22.7	83.2	0.36	5.0	5.60	5.49	3.42
Core #70 38–41	5/14/2010	<0.01	420	0.33	3,910	101	13.0	64.6	0.10	1.5	0.42	4.63	1.41
Core #70 50–52	5/14/2010	<0.01	405	0.36	3,580	93.6	12.8	69.3	0.20	1.6	0.74	3.79	1.26
					Snake River								
Core #73 0–2	5/19/2010	<0.01	373	0.36	2,700	76.4	11.3	44.2	0.36	2.4	0.14	6.69	1.36
Core #80 0–1	5/19/2010	<0.01	370	0.38	3,420	92.6	14.7	50.8	0.42	2.2	0.25	7.57	1.36
Core #81 0–3	5/19/2010	<0.01	372	0.42	3,970	98.7	14.1	53.2	0.40	2.3	0.13	7.82	1.42
Core #84 5–15	10/13/2010	0.039	361	0.35	3,190	88.5	14.0	49.0	0.10	2.2	0.11	4.22	1.10
Core #84 25–35	10/13/2010	0.030	344	0.36	3,010	86.0	13.0	49.1	0.22	2.5	0.11	4.63	1.17
Core #84 45–55	10/13/2010	0.031	388	0.41	2,850	80.9	12.7	47.5	0.20	2.8	0.10	4.73	1.21
Core #84 65–75	10/13/2010	0.029	324	0.47	2,640	71.4	13.1	54.0	0.38	3.1	0.11	5.88	1.59
Core #84 85–95	10/13/2010	0.069	342	0.40	2,470	69.5	11.9	72.4	0.22	2.3	0.10	4.76	1.21
Core #85 5–15	10/13/2010	0.075	341	0.39	2,500	70.9	11.6	45.2	0.75	2.4	0.16	4.80	1.13
Core #85 25–35	10/13/2010	0.035	323	0.35	2,830	82.5	12.9	47.5	0.28	2.5	0.12	5.95	1.11
Core #85 45–55	10/13/2010	0.027	348	0.35	3,470	105	15.1	54.7	0.25	2.5	0.12	4.38	1.21
Core #85 55–65	10/13/2010	0.026	329	0.34	3,020	87.4	13.3	48.6	0.20	2.4	0.11	5.42	1.17
Core #87 0–19	5/19/2010	<0.01	327	0.37	3,030	88.1	12.5	51.2	0.66	2.7	0.58	5.17	1.26
Core #88 0–3	5/19/2010	<0.01	380	0.44	3,410	84.7	14.6	72.5	0.47	2.9	0.15	7.65	1.44
Core #89 0–3	5/19/2010	<0.01	379	0.45	3,370	81.8	13.1	53.4	0.44	2.2	0.16	6.42	1.43

Appendix 4—Quality-Assurance and Quality-Control Data for Major and Trace Elements in Bed-Sediment Core Samples

Appendix 4.   Quality-assurance and quality-control data for major and trace elements in bed-sediment core samples collected in Lower Granite Reservoir and the Clearwater and Snake Rivers just above their confluence, 2010.

[Concentrations in micrograms per gram unless otherwise noted; #, number; <, less than lower reporting limit; %, percent; NIST, National Institute of Standards and Technology; --, not analyzed]

Type of quality-control sample	Sample identifier	Set number	Calcium	Magne-sium	Potassium	Sodium	Phospho-rus	Aluminum
Environmental	Core #31 164–167	11150	27,800	12,700	16,600	18,800	1,020	74,800
Laboratory replicate	Core #31 164–167 replicate	11150	27,400	12,600	16,200	18,600	1,010	75,000
Laboratory replicate	Core #31 164–167 replicate	11150	27,500	12,600	16,500	18,900	1,040	73,700
Blank	Blank	11150	<100	<6	<15	<25	<5	<50
Blank	Blank	11150	<100	<6	<15	44.2	<5	<50
Blank	Blank	11150	<100	<6	<15	<25	<5	<50
Blank	Blank	11150	<100	<6	<15	<25	<5	<50
Blank	Blank	11150	<100	<6	<15	<25	<5	<50
Blank	Blank	11150	<100	<6	<15	<25	<5	<50
Standard reference material	MAG-1 found	11150	10,600	17,900	30,900	28,100	693	84,200
Standard reference material	MAG-1 true (Potts and others, 1992)	11150	9,790	18,090	29,500	28,400	711	86,660
Standard reference material	Percent recovery	11150	108.3%	98.9%	104.7%	98.9%	97.5%	97.2%
Standard reference material	NIST 8704 found	11150	27,600	11,300	20,400	5,800	955	60,200
Standard reference material	NIST 8704 true (National Institute of Standards and Technology, 2008)	11150	26,410	12,000	20,010	5,530	--	61,000
Standard reference material	Percent recovery	11150	104.5%	94.2%	101.9%	104.9%	--	98.7%
Standard reference material	SCO-1 found	11150	19,800	15,500	23,100	6,800	869	71,800
Standard reference material	SCO-1 true (Potts and others, 1992)	11150	18,700	16,400	23,000	6,670	899	72,370
Standard reference material	Percent recovery	11150	105.9%	94.5%	100.4%	101.9%	96.7%	99.2%
Standard reference material	NIST 2709 found	11150	20,900	15,900	21,500	12,400	644	79,500
Standard reference material	NIST 2709 true (National Institute of Standards and Technology, 2003a)	11150	18,900	15,100	20,300	11,600	620	75,000
Standard reference material	Percent recovery	11150	110.6%	105.3%	105.9%	106.9%	103.9%	106.0%
Standard reference material	GSD-8 found	11150	2,290	2,040	36,300	4,590	180	61,900
Standard reference material	GSD-8 true (Potts and others, 1992)	11150	1,790	1,510	23,500	3,490	130	40,800
Standard reference material	Percent recovery	11150	127.9%	135.1%	154.5%	131.5%	138.5%	151.7%
Standard reference material	NIST-2711 found	11150	27,100	10,100	24,300	11,600	807	65,200
Standard reference material	NIST-2711 true (National Institute of Standards and Technology, 2003b)	11150	28,800	10,500	24,500	11,400	860	65,300
Standard reference material	Percent recovery	11150	94.1%	96.2%	99.2%	101.8%	93.8%	99.8%
Standard reference material	GSD-5 found	11150	39,500	5,510	18,200	2,580	596	81,900
Standard reference material	GSD-5 true (Potts and others, 1992)	11150	38,200	5,900	17,400	2,970	610	81,300
Standard reference material	Percent recovery	11150	103.4%	93.4%	104.6%	86.9%	97.7%	100.7%
Standard reference material	GSD-3 found	11150	1,550	3,950	20,900	2,130	602	66,600
Standard reference material	GSD-3 true (Potts and others, 1992)	11150	1,570	4,200	20,400	2,370	610	63,700
Standard reference material	Percent recovery	11150	98.7%	94.0%	102.5%	89.9%	98.7%	104.6%

Appendix 4. Quality-assurance and quality-control data for major and trace elements in bed-sediment core samples collected in Lower Granite Reservoir and the Clearwater and Snake Rivers just above their confluence, 2010.—Continued

[Concentrations in micrograms per gram unless otherwise noted; #, number; <, less than lower reporting limit; %, percent; NIST, National Institute of Standards and Technology; --, not analyzed]

Type of quality-control sample	Sample identifier	Set number	Calcium	Magne-sium	Potassium	Sodium	Phospho-rus	Aluminum
Standard reference material	MAG-1 found	11150	10,500	17,800	29,800	29,000	680	87,800
Standard reference material	MAG-1 true (Potts and others, 1992)	11150	9,790	18,090	29,500	28,400	711	86,660
Standard reference material	Percent recovery	11150	107.3%	98.4%	101.0%	102.1%	95.6%	101.3%
Standard reference material	NIST 8704 found	11150	26,400	11,600	19,500	5,910	917	62,000
Standard reference material	NIST 8704 true (National Institute of Standards and Technology, 2008)	11150	26,410	12,000	20,010	5,530	--	61,000
Standard reference material	Percent recovery	11150	100.0%	96.7%	97.5%	106.9%	--	101.6%
Environmental	Core #84 65–75	11637	15,000	6,690	21,900	19,500	400	60,300
Laboratory replicate	Core #84 65–75 replicate	11637	14,500	6,590	21,100	19,500	401	59,200
Blank	Blank	11637	<100	13.4	<15	<25	<5	<50
Blank	Blank	11637	<100	<6	<15	<25	7.6	<50
Blank	Blank	11637	<100	<6	<15	112	21	<50
Blank	Blank	11637	<100	<6	<15	<25	6	<50
Blank	Blank	11637	<100	<6	<15	<25	<5	<50
Standard reference material	MAG-1 found	11637	10,900	18,900	32,100	30,300	739	88,800
Standard reference material	MAG-1 true (Potts and others, 1992)	11637	9,790	18,090	29,500	28,400	711	86,660
Standard reference material	Percent recovery	11637	111.3%	104.5%	108.8%	106.7%	103.9%	102.5%
Standard reference material	NIST 8704 found	11637	29,400	12,900	21,900	6,340	1,040	66,100
Standard reference material	NIST 8704 true (National Institute of Standards and Technology, 2008)	11637	26,410	12,000	20,010	5,530	--	61,000
Standard reference material	Percent recovery	11637	111.3%	107.5%	109.4%	114.6%	--	108.4%
Standard reference material	SCO-1 found	11637	19,500	16,500	23,700	6,900	912	74,300
Standard reference material	SCO-1 true (Potts and others, 1992)	11637	18,700	16,400	23,000	6,670	899	72,370
Standard reference material	Percent recovery	11637	104.3%	100.6%	103.0%	103.4%	101.4%	102.7%
Standard reference material	NIST 2709 found	11637	21,200	16,100	21,900	12,400	670	80,700
Standard reference material	NIST 2709 true (National Institute of Standards and Technology, 2003a)	11637	18,900	15,100	20,300	11,600	620	75,000
Standard reference material	Percent recovery	11637	112.2%	106.6%	107.9%	106.9%	108.1%	107.6%
Standard reference material	GSD-8 found	11637	1,510	1,390	24,000	3,130	127	40,300
Standard reference material	GSD-8 true (Potts and others, 1992)	11637	1,790	1,510	23,500	3,490	130	40,800
Standard reference material	Percent recovery	11637	84.4%	92.1%	102.1%	89.7%	97.7%	98.8%
Standard reference material	NIST 2709 found	11637	19,000	14,500	19,700	11,300	595	71,500
Standard reference material	NIST 2709 true (National Institute of Standards and Technology, 2003a)	11637	18,900	15,100	20,300	11,600	620	75,000
Standard reference material	Percent recovery	11637	100.5%	96.0%	97.0%	97.4%	96.0%	95.3%

Appendix 4.   Quality-assurance and quality-control data for major and trace elements in bed-sediment core samples collected in Lower Granite Reservoir and the Clearwater and Snake Rivers just above their confluence, 2010.—Continued

[Concentrations in micrograms per gram unless otherwise noted; #, number; <, less than lower reporting limit; %, percent; NIST, National Institute of Standards and Technology; --, not analyzed]

Type of quality-control sample	Sample identifier	Set number	Barium	Beryllium	Bismuth	Cadmium	Cerium	Cesium
Environmental	Core #31 164–167	11150	723	1.5	5.90	0.27	58.1	3.2
Laboratory replicate	Core #31 164–167 replicate	11150	712	1.6	6.42	0.28	58.9	3.3
Laboratory replicate	Core #31 164–167 replicate	11150	715	1.5	5.97	0.27	58.9	3.2
Blank	Blank	11150	<0.25	<0.03	<0.06	<0.007	<0.1	0.003
Blank	Blank	11150	<0.25	<0.03	<0.06	<0.007	<0.1	0.008
Blank	Blank	11150	0.73	<0.03	1.27	<0.007	0.26	0.007
Blank	Blank	11150	<0.25	<0.03	<0.06	<0.007	<0.1	0.005
Blank	Blank	11150	<0.25	<0.03	<0.06	<0.007	<0.1	<0.003
Blank	Blank	11150	<0.25	0.03	<0.06	<0.007	<0.1	<0.003
Standard reference material	MAG-1 found	11150	489	2.5	9.15	0.19	75.6	8.8
Standard reference material	MAG-1 true (Potts and others, 1992)	11150	479	3.2	0.34	0.202	88	8.6
Standard reference material	Percent recovery	11150	102.1%	78.1%	2,691.2%	94.1%	85.9%	102.3%
Standard reference material	NIST 8704 found	11150	415	1.5	55.2	3.1	55.9	5.9
Standard reference material	NIST 8704 true (National Institute of Standards and Technology, 2008)	11150	413	--	--	2.94	66.5	5.83
Standard reference material	Percent recovery	11150	100.5%	--	--	105.4%	84.1%	101.2%
Standard reference material	SCO-1 found	11150	577	1.4	11.7	0.14	54.5	8.1
Standard reference material	SCO-1 true (Potts and others, 1992)	11150	570	1.84	0.37	0.14	62	7.8
Standard reference material	Percent recovery	11150	101.2%	76.1%	3,162.2%	100.0%	87.9%	103.8%
Standard reference material	NIST 2709 found	11150	985	3.1	7.18	0.4	46.4	6
Standard reference material	NIST 2709 true (National Institute of Standards and Technology, 2003a)	11150	968	--	--	0.38	42	5.3
Standard reference material	Percent recovery	11150	101.8%	--	--	105.3%	110.5%	113.2%
Standard reference material	GSD-8 found	11150	653	2.1	11.5	<0.007	77.5	5
Standard reference material	GSD-8 true (Potts and others, 1992)	11150	480	2	0.19	0.081	54	3.6
Standard reference material	Percent recovery	11150	136.0%	105.0%	6,052.6%	--	143.5%	138.9%
Standard reference material	NIST-2711 found	11150	726	1.9	415	40.5	73.2	7
Standard reference material	NIST-2711 true (National Institute of Standards and Technology, 2003b)	11150	726	--	--	41.7	69	6.1
Standard reference material	Percent recovery	11150	100.0%	--	--	97.1%	106.1%	114.8%
Standard reference material	GSD-5 found	11150	435	1.8	44	0.9	88.8	9.3
Standard reference material	GSD-5 true (Potts and others, 1992)	11150	440	2.3	2.4	0.82	89	9.4
Standard reference material	Percent recovery	11150	98.9%	78.3%	1,833.3%	109.8%	99.8%	98.9%
Standard reference material	GSD-3 found	11150	616	1.1	15.9	0.2	61.8	7.9
Standard reference material	GSD-3 true (Potts and others, 1992)	11150	615	1.5	0.79	0.1	64	7.8
Standard reference material	Percent recovery	11150	100.2%	73.3%	2,012.7%	200.0%	96.6%	101.3%

Appendix 4. Quality-assurance and quality-control data for major and trace elements in bed-sediment core samples collected in Lower Granite Reservoir and the Clearwater and Snake Rivers just above their confluence, 2010.—Continued

[Concentrations in micrograms per gram unless otherwise noted; #, number; <, less than lower reporting limit; %, percent; NIST, National Institute of Standards and Technology; --, not analyzed]

Type of quality-control sample	Sample identifier	Set number	Barium	Beryllium	Bismuth	Cadmium	Cerium	Cesium
Standard reference material	MAG-1 found	11150	489	2.7	9.53	0.19	74.9	9
Standard reference material	MAG-1 true (Potts and others, 1992)	11150	479	3.2	0.34	0.202	88	8.6
Standard reference material	Percent recovery	11150	102.1%	84.4%	2,802.9%	94.1%	85.1%	104.7%
Standard reference material	NIST 8704 found	11150	423	1.7	55.3	3	53.5	6.2
Standard reference material	NIST 8704 true (National Institute of Standards and Technology, 2008)	11150	413	--	--	2.94	66.5	5.83
Standard reference material	Percent recovery	11150	102.4%	--	--	102.0%	80.5%	106.3%
Environmental	Core #84 65–75	11637	1,040	1.8	10.4	0.02	34.7	2.5
Laboratory replicate	Core #84 65–75 replicate	11637	1,010	1.9	10.4	0.02	35.6	2.4
Blank	Blank	11637	<0.25	<0.03	<0.06	<0.007	<0.1	0.004
Blank	Blank	11637	<0.25	<0.03	<0.06	<0.007	<0.1	0.004
Blank	Blank	11637	0.27	<0.03	0.15	<0.007	<0.1	0.004
Blank	Blank	11637	<0.25	<0.03	<0.06	<0.007	<0.1	<0.003
Blank	Blank	11637	<0.25	<0.03	<0.06	<0.007	<0.1	<0.003
Standard reference material	MAG-1 found	11637	544	2.9	10.7	0.31	91.7	9.6
Standard reference material	MAG-1 true (Potts and others, 1992)	11637	479	3.20	0.34	0.20	88.0	8.60
Standard reference material	Percent recovery	11637	113.6%	90.6%	3,147.1%	153.5%	104.2%	111.6%
Standard reference material	NIST 8704 found	11637	481	1.8	60.6	3.4	65.1	6.8
Standard reference material	NIST 8704 true (National Institute of Standards and Technology, 2008)	11637	413	--	--	2.94	66.5	5.83
Standard reference material	Percent recovery	11637	116.5%	--	--	115.6%	97.9%	116.6%
Standard reference material	SCO-1 found	11637	628	1.6	12.4	0.15	60.1	8.7
Standard reference material	SCO-1 true (Potts and others, 1992)	11637	570	1.84	0.37	0.14	62.0	7.80
Standard reference material	Percent recovery	11637	110.2%	87.0%	3,351.4%	107.1%	96.9%	111.5%
Standard reference material	NIST 2709 found	11637	1,060	2.9	7.44	0.41	49.1	6.3
Standard reference material	NIST 2709 true (National Institute of Standards and Technology, 2003a)	11637	968	--	--	0.38	42.0	5.30
Standard reference material	Percent recovery	11637	109.5%	--	--	107.9%	116.9%	118.9%
Standard reference material	GSD-8 found	11637	484	1.6	8.15	0.03	55.2	3.6
Standard reference material	GSD-8 true (Potts and others, 1992)	11637	480	2.00	0.19	0.08	54.0	3.60
Standard reference material	Percent recovery	11637	100.8%	80.0%	4,289.5%	37.0%	102.2%	100.0%
Standard reference material	NIST 2709 found	11637	905	2.9	6.58	0.34	42	5.4
Standard reference material	NIST 2709 true (National Institute of Standards and Technology, 2003a)	11637	968	--	--	0.38	42.0	5.30
Standard reference material	Percent recovery	11637	93.5%	--	--	89.5%	100.0%	101.9%

Appendix 4.   Quality-assurance and quality-control data for major and trace elements in bed-sediment core samples collected in Lower Granite Reservoir and the Clearwater and Snake Rivers just above their confluence, 2010.—Continued

[Concentrations in micrograms per gram unless otherwise noted; #, number; <, less than lower reporting limit; %, percent; NIST, National Institute of Standards and Technology; --, not analyzed]

Type of quality-control sample	Sample identifier	Set number	Chro-mium	Cobalt	Copper	Gallium	Iron	Lantha-num	Lead
Environmental	Core #31 164–167	11150	52.4	20.9	36.7	19.8	45,000	30.0	15.5
Laboratory replicate	Core #31 164–167 replicate	11150	52.4	20.8	39.3	19.6	44,800	30.4	16.8
Laboratory replicate	Core #31 164–167 replicate	11150	52.9	20.7	36.6	19.8	45,300	30.1	15.8
Blank	Blank	11150	<0.5	<0.03	<2	<0.015	<50	<0.05	<0.4
Blank	Blank	11150	<0.5	<0.03	<2	<0.015	<50	<0.05	<0.4
Blank	Blank	11150	<0.5	<0.03	<2	<0.015	<50	0.16	3.35
Blank	Blank	11150	<0.5	<0.03	<2	<0.015	<50	<0.05	<0.4
Blank	Blank	11150	<0.5	<0.03	<2	<0.015	<50	<0.05	<0.4
Blank	Blank	11150	<0.5	<0.03	<2	<0.015	<50	<0.05	<0.4
Standard reference material	MAG-1 found	11150	104	23.1	30.2	24.5	49,100	36.8	24.3
Standard reference material	MAG-1 true (Potts and others, 1992)	11150	97	20.4	30	20.4	47,600	43	24
Standard reference material	Percent recovery	11150	107.2%	113.2%	100.7%	120.1%	103.2%	85.6%	101.3%
Standard reference material	NIST 8704 found	11150	121	14	91.4	16.7	40,200	26.6	146
Standard reference material	NIST 8704 true (National Institute of Standards and Technology, 2008)	11150	121.9	13.6	--	--	39,700	--	150
Standard reference material	Percent recovery	11150	99.3%	102.9%	--	--	101.3%	--	97.3%
Standard reference material	SCO-1 found	11150	71.9	11.9	32.4	18.4	35,800	27.6	31.2
Standard reference material	SCO-1 true (Potts and others, 1992)	11150	68	10.5	28.7	15	35,900	29.5	31
Standard reference material	Percent recovery	11150	105.7%	113.3%	112.9%	122.7%	99.7%	93.6%	100.6%
Standard reference material	NIST 2709 found	11150	122	14.4	37.8	18	36,900	23.5	18.7
Standard reference material	NIST 2709 true (National Institute of Standards and Technology, 2003a)	11150	130	13.4	34.6	14	35,000	23	18.9
Standard reference material	Percent recovery	11150	93.8%	107.5%	109.2%	128.6%	105.4%	102.2%	98.9%
Standard reference material	GSD-8 found	11150	7.9	5	8.8	16.7	22,200	37.9	30.9
Standard reference material	GSD-8 true (Potts and others, 1992)	11150	7.6	3.6	4.1	10.8	15,380	30	21
Standard reference material	Percent recovery	11150	103.9%	138.9%	214.6%	154.6%	144.3%	126.3%	147.1%
Standard reference material	NIST-2711 found	11150	43.1	10.1	112	17.2	28,400	38.4	1100
Standard reference material	NIST-2711 true (National Institute of Standards and Technology, 2003b)	11150	47	10	114	15	28,900	40	1162
Standard reference material	Percent recovery	11150	91.7%	101.0%	98.2%	114.7%	98.3%	96.0%	94.7%
Standard reference material	GSD-5 found	11150	69.1	20.3	140	22.4	41,200	39.4	115
Standard reference material	GSD-5 true (Potts and others, 1992)	11150	70	18.9	137	20.3	41,000	46	112
Standard reference material	Percent recovery	11150	98.7%	107.4%	102.2%	110.3%	100.5%	85.7%	102.7%
Standard reference material	GSD-3 found	11150	80.2	11.6	183	16.7	45,900	32.7	42
Standard reference material	GSD-3 true (Potts and others, 1992)	11150	87	11.7	177	15.9	45,500	39	40
Standard reference material	Percent recovery	11150	92.2%	99.1%	103.4%	105.0%	100.9%	83.8%	105.0%

Appendix 4.   Quality-assurance and quality-control data for major and trace elements in bed-sediment core samples collected in Lower Granite Reservoir and the Clearwater and Snake Rivers just above their confluence, 2010.—Continued

[Concentrations in micrograms per gram unless otherwise noted; #, number; <, less than lower reporting limit; %, percent; NIST, National Institute of Standards and Technology; --, not analyzed]

Type of quality-control sample	Sample identifier	Set number	Chro-mium	Cobalt	Copper	Gallium	Iron	Lantha-num	Lead
Standard reference material	MAG-1 found	11150	97.6	22	29.7	23.6	47,600	37.4	24.6
Standard reference material	MAG-1 true (Potts and others, 1992)	11150	97	20.4	30	20.4	47,600	43	24
Standard reference material	Percent recovery	11150	100.6%	107.8%	99.0%	115.7%	100.0%	87.0%	102.5%
Standard reference material	NIST 8704 found	11150	115	13.6	89.7	16.2	38,600	25.7	145
Standard reference material	NIST 8704 true (National Institute of Standards and Technology, 2008)	11150	121.9	13.6	--	--	39,700	--	150
Standard reference material	Percent recovery	11150	94.3%	100.0%	--	--	97.2%	--	96.7%
Environmental	Core #84 65–75	11637	22.9	11.1	18.9	16.3	22,300	18.8	28.6
Laboratory replicate	Core #84 65–75 replicate	11637	21.3	10.9	19.7	16.1	21,700	19.4	28.7
Blank	Blank	11637	<0.5	<0.03	<2	<0.015	<50	<0.05	<0.4
Blank	Blank	11637	<0.5	<0.03	<2	<0.015	<50	<0.05	<0.4
Blank	Blank	11637	<0.5	<0.03	<2	<0.015	<50	<0.05	0.41
Blank	Blank	11637	<0.5	<0.03	<2	<0.015	<50	<0.05	<0.4
Blank	Blank	11637	<0.5	<0.03	<2	0.02	<50	<0.05	<0.4
Standard reference material	MAG-1 found	11637	107	23.8	32	25.9	51,300	44	28.3
Standard reference material	MAG-1 true (Potts and others, 1992)	11637	97	20.4	30.0	20.4	47,600	43.0	24.0
Standard reference material	Percent recovery	11637	110.3%	116.7%	106.7%	127.0%	107.8%	102.3%	117.9%
Standard reference material	NIST 8704 found	11637	127	15	98.1	18.1	43,600	31.1	162
Standard reference material	NIST 8704 true (National Institute of Standards and Technology, 2008)	11637	122	13.6	--	--	39,700	--	150.0
Standard reference material	Percent recovery	11637	104.2%	110.3%	--	--	109.8%	--	108.0%
Standard reference material	SCO-1 found	11637	72.2	12.1	30.2	19	37,100	30.3	33.1
Standard reference material	SCO-1 true (Potts and others, 1992)	11637	68.0	10.5	28.7	15.0	35,900	29.5	31.0
Standard reference material	Percent recovery	11637	106.2%	115.2%	105.2%	126.7%	103.3%	102.7%	106.8%
Standard reference material	NIST 2709 found	11637	120	14.8	38.8	19.2	38,300	24.3	19.9
Standard reference material	NIST 2709 true (National Institute of Standards and Technology, 2003a)	11637	130	13.4	34.6	14.0	35,000	23.0	18.9
Standard reference material	Percent recovery	11637	92.3%	110.4%	112.1%	137.1%	109.4%	105.7%	105.3%
Standard reference material	GSD-8 found	11637	9.7	3.4	5	11.1	15,000	26.7	22
Standard reference material	GSD-8 true (Potts and others, 1992)	11637	7.60	3.6	4.1	10.8	15,380	30.0	21.0
Standard reference material	Percent recovery	11637	127.6%	94.4%	122.0%	102.8%	97.5%	89.0%	104.8%
Standard reference material	NIST 2709 found	11637	104	12.7	33.4	16.1	33,800	21	17.5
Standard reference material	NIST 2709 true (National Institute of Standards and Technology, 2003a)	11637	130	13.4	34.6	14.0	35,000	23.0	18.9
Standard reference material	Percent recovery	11637	80.0%	94.8%	96.5%	115.0%	96.6%	91.3%	92.6%

Appendix 4.   Quality-assurance and quality-control data for major and trace elements in bed-sediment core samples collected in Lower Granite Reservoir and the Clearwater and Snake Rivers just above their confluence, 2010.—Continued

[Concentrations in micrograms per gram unless otherwise noted; #, number; <, less than lower reporting limit; %, percent; NIST, National Institute of Standards and Technology; --, not analyzed]

Type of quality-control sample	Sample identifier	Set number	Lithium	Manga-nese	Molybde-num	Nickel	Niobium	Rubidium
Environmental	Core #31 164–167	11150	20.7	790	0.86	26.2	14	70.8
Laboratory replicate	Core #31 164–167 replicate	11150	22.0	798	0.88	25.6	14	70.2
Laboratory replicate	Core #31 164–167 replicate	11150	21.3	806	0.91	25.9	14	70.4
Blank	Blank	11150	<0.3	<0.7	<0.05	<0.3	0.2	<0.014
Blank	Blank	11150	0.3	<0.7	<0.05	<0.3	0.7	<0.014
Blank	Blank	11150	<0.3	<0.7	<0.05	<0.3	0.48	<0.014
Blank	Blank	11150	<0.3	<0.7	<0.05	<0.3	0.34	<0.014
Blank	Blank	11150	<0.3	<0.7	<0.05	<0.3	0.1	<0.014
Blank	Blank	11150	<0.3	<0.7	<0.05	<0.3	<0.1	<0.014
Standard reference material	MAG-1 found	11150	74.3	780	1.1	51.9	14	158
Standard reference material	MAG-1 true (Potts and others, 1992)	11150	79	760	1.6	53	12	149
Standard reference material	Percent recovery	11150	94.1%	102.6%	68.8%	97.9%	116.7%	106.0%
Standard reference material	NIST 8704 found	11150	44.3	585	4.6	43.2	7.6	107
Standard reference material	NIST 8704 true (National Institute of Standards and Technology, 2008)	11150	--	544	--	42.9	--	--
Standard reference material	Percent recovery	11150	--	107.5%	--	100.7%	--	--
Standard reference material	SCO-1 found	11150	42	409	1.3	27.7	10	119
Standard reference material	SCO-1 true (Potts and others, 1992)	11150	45	410	1.37	27	11	112
Standard reference material	Percent recovery	11150	93.3%	99.8%	94.9%	102.6%	90.9%	106.3%
Standard reference material	NIST 2709 found	11150	55.5	594	2.2	88	11	103
Standard reference material	NIST 2709 true (National Institute of Standards and Technology, 2003a)	11150	--	538	2	88	--	96
Standard reference material	Percent recovery	11150	--	110.4%	110.0%	100.0%	--	107.3%
Standard reference material	GSD-8 found	11150	17.6	510	0.82	2.4	51	198
Standard reference material	GSD-8 true (Potts and others, 1992)	11150	13.2	310	0.54	2.7	35	132
Standard reference material	Percent recovery	11150	133.3%	164.5%	151.9%	88.9%	145.7%	150.0%
Standard reference material	NIST-2711 found	11150	26.5	655	1.7	20	16	120
Standard reference material	NIST-2711 true (National Institute of Standards and Technology, 2003b)	11150	--	638	1.6	20.6	--	110
Standard reference material	Percent recovery	11150	--	102.7%	106.3%	97.1%	--	109.1%
Standard reference material	GSD-5 found	11150	43.3	1,220	1.3	37.6	11	128
Standard reference material	GSD-5 true (Potts and others, 1992)	11150	45	1,160	1.2	34	19	118
Standard reference material	Percent recovery	11150	96.2%	105.2%	108.3%	110.6%	57.9%	108.5%
Standard reference material	GSD-3 found	11150	32.4	424	96.9	27.2	6.8	83.2
Standard reference material	GSD-3 true (Potts and others, 1992)	11150	33	390	92	25.6	16	79
Standard reference material	Percent recovery	11150	98.2%	108.7%	105.3%	106.3%	42.5%	105.3%

Appendix 4. Quality-assurance and quality-control data for major and trace elements in bed-sediment core samples collected in Lower Granite Reservoir and the Clearwater and Snake Rivers just above their confluence, 2010.—Continued

[Concentrations in micrograms per gram unless otherwise noted; #, number; <, less than lower reporting limit; %, percent; NIST, National Institute of Standards and Technology; --, not analyzed]

Type of quality-control sample	Sample identifier	Set number	Lithium	Manganese	Molybdenum	Nickel	Niobium	Rubidium
Standard reference material	MAG-1 found	11150	74.8	743	1.1	50.5	12	149
Standard reference material	MAG-1 true (Potts and others, 1992)	11150	79	760	1.6	53	12	149
Standard reference material	Percent recovery	11150	94.7%	97.8%	68.8%	95.3%	100.0%	100.0%
Standard reference material	NIST 8704 found	11150	44.3	559	4.5	42.6	7	102
Standard reference material	NIST 8704 true (National Institute of Standards and Technology, 2008)	11150	--	544	--	42.9	--	--
Standard reference material	Percent recovery	11150	--	102.8%	--	99.3%	--	--
Environmental	Core #84 65–75	11637	0.02	389	0.47	12.7	12	92.3
Laboratory replicate	Core #84 65–75 replicate	11637	--	375	0.47	12	12	91.7
Blank	Blank	11637	<0.3	2.3	<0.05	<0.3	<0.1	<0.014
Blank	Blank	11637	<0.3	<0.7	<0.05	<0.3	2	<0.014
Blank	Blank	11637	<0.3	<0.7	<0.05	<0.3	0.58	<0.014
Blank	Blank	11637	<0.3	<0.7	<0.05	<0.3	0.31	<0.014
Blank	Blank	11637	0.8	<0.7	<0.05	<0.3	0.2	<0.014
Standard reference material	MAG-1 found	11637	79.2	809	1.2	54.5	15	161
Standard reference material	MAG-1 true (Potts and others, 1992)	11637	79.0	760	1.60	53.0	12.0	149
Standard reference material	Percent recovery	11637	100.3%	106.4%	75.0%	102.8%	125.0%	108.1%
Standard reference material	NIST 8704 found	11637	48	625	4.1	47.8	9.7	114
Standard reference material	NIST 8704 true (National Institute of Standards and Technology, 2008)	11637	--	544	--	42.9	--	--
Standard reference material	Percent recovery	11637	--	114.9%	--	111.4%	--	--
Standard reference material	SCO-1 found	11637	44.9	413	1.2	27.9	10	120
Standard reference material	SCO-1 true (Potts and others, 1992)	11637	45.0	410	1.37	27.0	11.0	112
Standard reference material	Percent recovery	11637	99.8%	100.7%	87.6%	103.3%	90.9%	107.1%
Standard reference material	NIST 2709 found	11637	56.3	610	2.2	90	9.4	105
Standard reference material	NIST 2709 true (National Institute of Standards and Technology, 2003a)	11637	--	538	2	88.0	--	96
Standard reference material	Percent recovery	11637	--	113.4%	110.0%	102.3%	--	109.4%
Standard reference material	GSD-8 found	11637	11.8	336	0.56	1.6	33	135
Standard reference material	GSD-8 true (Potts and others, 1992)	11637	13.2	310	0.54	2.7	35.0	132
Standard reference material	Percent recovery	11637	89.4%	108.4%	103.7%	59.3%	94.3%	102.3%
Standard reference material	NIST 2709 found	11637	52	540	1.9	77.8	8.9	90.7
Standard reference material	NIST 2709 true (National Institute of Standards and Technology, 2003a)	11637	--	538	2	88.0	--	96
Standard reference material	Percent recovery	11637	--	100.4%	95.0%	88.4%	--	94.5%

Appendix 4.   Quality-assurance and quality-control data for major and trace elements in bed-sediment core samples collected in Lower Granite Reservoir and the Clearwater and Snake Rivers just above their confluence, 2010.—Continued

[Concentrations in micrograms per gram unless otherwise noted; #, number; <, less than lower reporting limit; %, percent; NIST, National Institute of Standards and Technology; --, not analyzed]

Type of quality-control sample	Sample identifier	Set number	Scandium	Silver	Strontium	Thallium	Titanium	Vanadium
Environmental	Core #31 164–167	11150	17.5	<0.01	349	0.40	6,170	160
Laboratory replicate	Core #31 164–167 replicate	11150	17.3	<0.01	343	0.40	6,090	158
Laboratory replicate	Core #31 164–167 replicate	11150	17.5	<0.01	347	0.41	6,180	161
Blank	Blank	11150	<0.04	12.3	<0.8	<0.08	<40	<0.15
Blank	Blank	11150	<0.04	<0.01	<0.8	<0.08	<40	<0.15
Blank	Blank	11150	<0.04	<0.01	<0.8	<0.08	<40	<0.15
Blank	Blank	11150	<0.04	<0.01	<0.8	<0.08	<40	<0.15
Blank	Blank	11150	<0.04	2.17	<0.8	<0.08	<40	<0.15
Blank	Blank	11150	<0.04	0.816	<0.8	<0.08	<40	<0.15
Standard reference material	MAG-1 found	11150	18	<0.01	148	0.71	3,400	153
Standard reference material	MAG-1 true (Potts and others, 1992)	11150	17.2	0.08	146	0.59	4,500	140
Standard reference material	Percent recovery	11150	104.7%	--	101.4%	120.3%	75.6%	109.3%
Standard reference material	NIST 8704 found	11150	11.4	<0.01	137	1.06	2,200	97.8
Standard reference material	NIST 8704 true (National Institute of Standards and Technology, 2008)	11150	11.26	--	--	--	4,570	94.6
Standard reference material	Percent recovery	11150	101.2%	--	--	--	48.1%	103.4%
Standard reference material	SCO-1 found	11150	11.9	<0.01	175	0.69	2,880	145
Standard reference material	SCO-1 true (Potts and others, 1992)	11150	10.8	0.134	174	0.72	3,760	131
Standard reference material	Percent recovery	11150	110.2%	--	100.6%	95.8%	76.6%	110.7%
Standard reference material	NIST 2709 found	11150	12.8	<0.01	248	0.68	3,250	127
Standard reference material	NIST 2709 true (National Institute of Standards and Technology, 2003a)	11150	12	--	231	0.74	3,420	112
Standard reference material	Percent recovery	11150	106.7%	--	107.4%	91.9%	95.0%	113.4%
Standard reference material	GSD-8 found	11150	7.5	<0.01	74.9	1.08	4,270	36
Standard reference material	GSD-8 true (Potts and others, 1992)	11150	5.7	0.062	52	0.78	3,660	26
Standard reference material	Percent recovery	11150	131.6%	--	144.0%	138.5%	116.7%	138.5%
Standard reference material	NIST-2711 found	11150	9.3	2.64	253	2.55	2,300	85.8
Standard reference material	NIST-2711 true (National Institute of Standards and Technology, 2003b)	11150	9	4.63	245.3	2.47	3,060	81.6
Standard reference material	Percent recovery	11150	103.3%	57.0%	103.1%	103.2%	75.2%	105.1%
Standard reference material	GSD-5 found	11150	13.9	<0.01	219	1.19	3,370	114
Standard reference material	GSD-5 true (Potts and others, 1992)	11150	14.5	0.36	204	1.16	5,400	109
Standard reference material	Percent recovery	11150	95.9%	--	107.4%	102.6%	62.4%	104.6%
Standard reference material	GSD-3 found	11150	12.9	<0.01	93.7	0.48	3,200	124
Standard reference material	GSD-3 true (Potts and others, 1992)	11150	14.3	0.59	90	0.58	6,360	120
Standard reference material	Percent recovery	11150	90.2%	--	104.1%	82.8%	50.3%	103.3%

Appendix 4. Quality-assurance and quality-control data for major and trace elements in bed-sediment core samples collected in Lower Granite Reservoir and the Clearwater and Snake Rivers just above their confluence, 2010.—Continued

[Concentrations in micrograms per gram unless otherwise noted; #, number; <, less than lower reporting limit; %, percent; NIST, National Institute of Standards and Technology; --, not analyzed]

Type of quality-control sample	Sample identifier	Set number	Scandium	Silver	Strontium	Thallium	Titanium	Vanadium
Standard reference material	MAG-1 found	11150	17.5	<0.01	141	0.73	3,150	145
Standard reference material	MAG-1 true (Potts and others, 1992)	11150	17.2	0.08	146	0.59	4,500	140
Standard reference material	Percent recovery	11150	101.7%	--	96.6%	123.7%	70.0%	103.6%
Standard reference material	NIST 8704 found	11150	11.6	<0.01	133	1.06	2,080	94.9
Standard reference material	NIST 8704 true (National Institute of Standards and Technology, 2008)	11150	11.26	--	--	--	4,570	94.6
Standard reference material	Percent recovery	11150	103.0%	--	--	--	45.5%	100.3%
Environmental	Core #84 65–75	11637	7.6	0.029	324	0.47	2,640	71.4
Laboratory replicate	Core #84 65–75 replicate	11637	7.4	0.039	323	0.47	2,580	69.9
Blank	Blank	11637	<0.04	0.029	<0.8	<0.08	<40	<0.15
Blank	Blank	11637	<0.04	<0.01	<0.8	<0.08	<40	<0.15
Blank	Blank	11637	<0.04	<0.01	<0.8	<0.08	<40	<0.15
Blank	Blank	11637	<0.04	<0.01	<0.8	<0.08	<40	<0.15
Blank	Blank	11637	<0.04	<0.01	<0.8	<0.08	<40	<0.15
Standard reference material	MAG-1 found	11637	18.6	0.065	155	0.75	3,640	155
Standard reference material	MAG-1 true (Potts and others, 1992)	11637	17.2	0.08	146	0.59	4,500	140
Standard reference material	Percent recovery	11637	108.1%	81.3%	106.2%	127.1%	80.9%	110.7%
Standard reference material	NIST 8704 found	11637	12.8	0.317	152	1.14	2,670	103
Standard reference material	NIST 8704 true (National Institute of Standards and Technology, 2008)	11637	11.3	--	--	--	4,570	94.6
Standard reference material	Percent recovery	11637	113.7%	--	--	--	58.4%	108.9%
Standard reference material	SCO-1 found	11637	12.5	0.103	181	0.72	2,850	143
Standard reference material	SCO-1 true (Potts and others, 1992)	11637	10.8	0.134	174	0.72	3,760	131
Standard reference material	Percent recovery	11637	115.7%	76.9%	104.0%	100.0%	75.8%	109.2%
Standard reference material	NIST 2709 found	11637	13.3	0.339	255	0.66	3,190	126
Standard reference material	NIST 2709 true (National Institute of Standards and Technology, 2003a)	11637	12.0	--	231	0.74	3,420	112.0
Standard reference material	Percent recovery	11637	110.8%	--	110.4%	89.2%	93.3%	112.5%
Standard reference material	GSD-8 found	11637	5	<0.01	51.4	0.72	2,750	24.6
Standard reference material	GSD-8 true (Potts and others, 1992)	11637	5.7	0.062	52.0	0.78	3,660	26
Standard reference material	Percent recovery	11637	87.7%	--	98.8%	92.3%	75.1%	94.6%
Standard reference material	NIST 2709 found	11637	11.3	0.306	218	0.65	2,840	109
Standard reference material	NIST 2709 true (National Institute of Standards and Technology, 2003a)	11637	12.0	--	231	0.74	3,420	112.0
Standard reference material	Percent recovery	11637	94.2%	--	94.4%	87.8%	83.0%	97.3%

Appendix 4. Quality-assurance and quality-control data for major and trace elements in bed-sediment core samples collected in Lower Granite Reservoir and the Clearwater and Snake Rivers just above their confluence, 2010.—Continued

[Concentrations in micrograms per gram unless otherwise noted; #, number; <, less than lower reporting limit; %, percent; NIST, National Institute of Standards and Technology; --, not analyzed]

Type of quality-control sample	Sample identifier	Set number	Yttrium	Zinc	Antimony	Arsenic	Thorium	Uranium
Environmental	Core #31 164–167	11150	25.2	102	0.81	6.1	7.62	3.73
Laboratory replicate	Core #31 164–167 replicate	11150	24.7	102	2.0	6.2	8.36	3.81
Laboratory replicate	Core #31 164–167 replicate	11150	25.0	103	0.77	6.0	7.82	3.85
Blank	Blank	11150	<0.05	<3	0.32	<1	<0.1	<0.02
Blank	Blank	11150	<0.05	<3	<0.04	<1	<0.1	<0.02
Blank	Blank	11150	<0.05	<3	<0.04	<1	<0.1	<0.02
Blank	Blank	11150	<0.05	<3	<0.04	<1	<0.1	<0.02
Blank	Blank	11150	<0.05	<3	0.23	<1	<0.1	<0.02
Blank	Blank	11150	<0.05	<3	0.22	<1	<0.1	<0.02
Standard reference material	MAG-1 found	11150	19.8	136	1	9.8	11	2.47
Standard reference material	MAG-1 true (Potts and others, 1992)	11150	28	130	0.96	9.2	11.9	2.7
Standard reference material	Percent recovery	11150	70.7%	104.6%	104.2%	106.5%	92.4%	91.5%
Standard reference material	NIST 8704 found	11150	20	393	2.9	17	8.44	2.72
Standard reference material	NIST 8704 true (National Institute of Standards and Technology, 2008)	11150	--	408	3.07	--	9.07	3.09
Standard reference material	Percent recovery	11150	--	96.3%	94.5%	--	93.1%	88.0%
Standard reference material	SCO-1 found	11150	17.8	107	2.6	12.7	8.92	2.7
Standard reference material	SCO-1 true (Potts and others, 1992)	11150	26	103	2.5	12.4	9.7	3
Standard reference material	Percent recovery	11150	68.5%	103.9%	104.0%	102.4%	92.0%	90.0%
Standard reference material	NIST 2709 found	11150	16.2	111	7.4	19.2	11.1	3.06
Standard reference material	NIST 2709 true (National Institute of Standards and Technology, 2003a)	11150	18	106	7.9	17.7	11	3
Standard reference material	Percent recovery	11150	90.0%	104.7%	93.7%	108.5%	100.9%	102.0%
Standard reference material	GSD-8 found	11150	19.2	65.9	0.39	4	19.5	4.37
Standard reference material	GSD-8 true (Potts and others, 1992)	11150	18	43	0.24	2.4	13.4	3
Standard reference material	Percent recovery	11150	106.7%	153.3%	162.5%	166.7%	145.5%	145.7%
Standard reference material	NIST-2711 found	11150	24.7	340	19.6	103	13.5	2.49
Standard reference material	NIST-2711 true (National Institute of Standards and Technology, 2003b)	11150	25	350.4	19.4	105	14	2.6
Standard reference material	Percent recovery	11150	98.8%	97.0%	101.0%	98.1%	96.4%	95.8%
Standard reference material	GSD-5 found	11150	16.9	257	4.2	78.1	14.7	2.37
Standard reference material	GSD-5 true (Potts and others, 1992)	11150	26	243	3.9	75	15.2	2.6
Standard reference material	Percent recovery	11150	65.0%	105.8%	107.7%	104.1%	96.7%	91.2%
Standard reference material	GSD-3 found	11150	12	52.6	6.5	18.6	8.54	1.43
Standard reference material	GSD-3 true (Potts and others, 1992)	11150	22	52	5.4	17.6	9.2	1.86
Standard reference material	Percent recovery	11150	54.5%	101.2%	120.4%	105.7%	92.8%	76.9%

Appendix 4. Quality-assurance and quality-control data for major and trace elements in bed-sediment core samples collected in Lower Granite Reservoir and the Clearwater and Snake Rivers just above their confluence, 2010.—Continued

[Concentrations in micrograms per gram unless otherwise noted; #, number; <, less than lower reporting limit; %, percent; NIST, National Institute of Standards and Technology; --, not analyzed]

Type of quality-control sample	Sample identifier	Set number	Yttrium	Zinc	Antimony	Arsenic	Thorium	Uranium
Standard reference material	MAG-1 found	11150	19.2	132	1	9.2	11.5	2.63
Standard reference material	MAG-1 true (Potts and others, 1992)	11150	28	130	0.96	9.2	11.9	2.7
Standard reference material	Percent recovery	11150	68.6%	101.5%	104.2%	100.0%	96.6%	97.4%
Standard reference material	NIST 8704 found	11150	19.6	385	2.9	16.5	8.3	2.81
Standard reference material	NIST 8704 true (National Institute of Standards and Technology, 2008)	11150	--	408	3.07	--	9.07	3.09
Standard reference material	Percent recovery	11150	--	94.4%	94.5%	--	91.5%	90.9%
Environmental	Core #84 65–75	11637	13.1	54.0	0.38	3.1	5.88	1.59
Laboratory replicate	Core #84 65–75 replicate	11637	12.2	53.4	0.27	3.0	5.93	1.44
Blank	Blank	11637	<0.05	<3	0.2	<1	<0.1	<0.02
Blank	Blank	11637	<0.05	<3	0.05	<1	<0.1	<0.02
Blank	Blank	11637	<0.05	<3	<0.04	<1	<0.1	<0.02
Blank	Blank	11637	<0.05	<3	<0.04	<1	<0.1	<0.02
Blank	Blank	11637	<0.05	<3	0.34	<1	<0.1	<0.02
Standard reference material	MAG-1 found	11637	22.2	142	0.91	9.9	12.9	2.92
Standard reference material	MAG-1 true (Potts and others, 1992)	11637	28.0	130	0.96	9.2	11.9	2.70
Standard reference material	Percent recovery	11637	79.3%	109.2%	94.8%	107.6%	108.4%	108.1%
Standard reference material	NIST 8704 found	11637	22.8	431	3.4	18.2	9.57	3.14
Standard reference material	NIST 8704 true (National Institute of Standards and Technology, 2008)	11637	--	408	3.07	--	9.1	3.09
Standard reference material	Percent recovery	11637	--	105.6%	110.7%	--	105.5%	101.6%
Standard reference material	SCO-1 found	11637	18.7	108	2.6	12.8	10.1	3.02
Standard reference material	SCO-1 true (Potts and others, 1992)	11637	26.0	103	2.50	12.4	9.7	3.00
Standard reference material	Percent recovery	11637	71.9%	104.9%	104.0%	103.2%	104.1%	100.7%
Standard reference material	NIST 2709 found	11637	16.6	114	7.1	19.4	12.1	3.12
Standard reference material	NIST 2709 true (National Institute of Standards and Technology, 2003a)	11637	18.0	106	7.90	17.7	11.0	3.00
Standard reference material	Percent recovery	11637	92.2%	107.5%	89.9%	109.6%	110.0%	104.0%
Standard reference material	GSD-8 found	11637	13.3	47	0.05	2.7	13.5	3.05
Standard reference material	GSD-8 true (Potts and others, 1992)	11637	18.0	43.0	0.24	2.4	13.4	3.00
Standard reference material	Percent recovery	11637	73.9%	109.3%	20.8%	112.5%	100.7%	101.7%
Standard reference material	NIST 2709 found	11637	14.3	99.3	6.7	16.8	10.4	2.78
Standard reference material	NIST 2709 true (National Institute of Standards and Technology, 2003a)	11637	18.0	106	7.90	17.7	11.0	3.00
Standard reference material	Percent recovery	11637	79.4%	93.7%	84.8%	94.9%	94.5%	92.7%

Publishing support provided by
Lafayette Publishing Service Center

Information regarding water resources in Texas is available at
http://tx.usgs.gov/

Braun and others—Grain-Size Distribution and Selected Major and Trace Element Concentrations in Bed-Sediment Cores—SIR 2012–5219

I SBN 978- 1- 4113-3512- 7

9 781411 335127